T0220416

Landprints

Reflections on place

and landscape

George Seddon

Foreword by Sir Gustav Nossal

CAMBRIDGE
UNIVERSITY PRESS

CAMBRIDGE UNIVERSITY PRESS
Cambridge, New York, Melbourne, Madrid, Cape Town,
Singapore, São Paulo, Delhi, Tokyo, Mexico City

Cambridge University Press
The Edinburgh Building, Cambridge CB2 8RU, UK

Published in the United States of America by Cambridge University Press, New York

www.cambridge.org
Information on this title: www.cambridge.org/9780521659994

© George Seddon 1997

This publication is in copyright. Subject to statutory exception
and to the provisions of relevant collective licensing agreements,
no reproduction of any part may take place without the written
permission of Cambridge University Press.

First published 1997
First paperback edition 1998
Re-issued 2011

A catalogue record for this publication is available from the British Library

Library of Congress Cataloguing in Publication data

Seddon, George, 1927– .
Landprints: reflections on place and landscape/George Seddon.
      p.   cm.
Includes bibliographical references (p.   ) and index.
ISBN 0-521-65999-x. (pb: alk. paper)
1. Natural history. 2. Landscape architecture. 3. Natural
history – Australia. 4. Landscape architecture – Australia.
1. Title.
QH45.2.S434   1997   97–27550
508.94–dc21

ISBN 978-0-521-58501-9 Hardback
ISBN 978-0-521-65999-4 Paperback

Additional resources for this publication at www.cambridge.org/9780521659994

Cambridge University Press has no responsibility for the persistence or
accuracy of URLs for external or third-party internet websites referred to in
this publication, and does not guarantee that any content on such websites is,
or will remain, accurate or appropriate.

Contents

List of illustrations   v
Foreword by Sir Gustav Nossal   vii
Acknowledgments   ix
Prelude: Dual allegiances   xi

FUGUE FOR SIX VOICES   1

Theme I   Talking: the language of landscape   5

1   The nature of Nature   7
2   Words and weeds   15
3   Journeys through a landscape   28
4   On *The Road to Botany Bay*   35
5   A Snowy River reader   49

Theme II   Perceiving: the eyes and the mind   61

6   The evolution of perceptual attitudes   64
7   Eurocentrism and Australian science   73
8   Figures in the landscape   83
9   Dreaming up a rainforest   90
10   Home thoughts from abroad   99

Theme III   Locating: the sense of place   105

11   Sense of place   109
12   The *genius loci* and the Australian landscape   113
13   Cuddlepie and other surrogates   119
14   Jet-set and parish pump   127
15   Placing the debate   136

Theme IV   Making: creating gardens and the evolution of styles   145

16   The suburban garden in Australia   149
17   The Australian backyard   153
18   Gardening across Australia   164
19   The garden as Paradise   176

Theme V   Analysing: ideologies and attitudes   187

20   The rhetoric and ethics of the environmental protest movement   189
21   The perfectibility of Nature   200

Theme VI   Sharing and caring: ecological frameworks   209

22   Biological pollution   211
23   The lie of the land   221
24   Eating the future   232
25   Felling the 'Groves of Life'   241

Coda   Learning to be at home: 'and then came Venice'   246

Index   249

Illustrations

Theme openings

The signature tune of Australia is its distinctive trees, so each of the six themes begins with a portrait.

Theme I   Snowgums (*Eucalyptus pauciflora*) on Mt Ginini in the ACT, tenacious survivors. COLIN TOTTERDELL   4

Theme II   Lowland coastal rainforest in North Queensland: *Palaquium galatoxylon* displays its sinuous and slender plank-buttresses, serving for support and gas exchange. COLIN TOTTERDELL   60

Theme III   *Xanthorrhoea preissii* south of Fremantle, WA; at first the grasstrees were classified as members of the Lily family, but they now have one of their own, the Xanthorrhoeaceae. MICHAL LEWI   104

Theme IV   *Eucalyptus rubida*, the candlebark: a handsome tree of the south-eastern Tablelands with a bark that colours in autumn – hence the specific name. COLIN TOTTERDELL   144

Theme V   *Eucalyptus wandoo* on land cleared for grazing on the Pinjarra Plain south of Perth. Like the mountain ash (*E. regnans*), wandoo regenerates best on an ash bed after fire. As with jarrah, the young seedling puts its growth effort underground for up to ten years, by developing a strong lignotuber and root system, before it becomes a sapling. The wood of wandoo is incredibly hard, and it is chemically inert in contact with steel bolts, so it is highly valued for construction. MICHAL LEWI   186

Theme VI   Rainforest edge at Lake Barrine on the Atherton Tableland in North Queensland. The margins of the forest are sealed off from light penetration by a luxuriant mantle of lianes – creepers, climbers and epiphytes. COLIN TOTTERDELL   208

Black and white

CSIRO research on kangaroo grass (*Themeda australis*) near Canberra, ACT.   17
Low-lying sandy coastline north of Perth, WA.   29
'The altar of the gods': the summit of Bogong.   58
Perth from Mt Eliza, 1983.   85
View from Mt Eliza, 1842.   85
Queen Victoria's view from Kings Park, 1911.   85
Rainforest: the Daintree, Far North Queensland: a collage by Jeannie Baker.   95
A whitewashed timber bridge over the Ovens River: 'Right for Bright'?   107

The bush newspaper: the scribbly gum, *Eucalyptus haemastoma*.   124
Moreton Bay figs breathing the very spirit of place at Rottnest.   140
A backyard shows its functions.   154
Two intricate natural geometries.   201
Australia Felix: the pastoral idyll.   216
Australia Infelix: the pastoral reality in too much of Australia.   217
The Groves of Life: *Araucaria bidwillii*, the bunya pine of south-eastern
     Queensland.   243
Oh brave New World: clearfelling overprints the land.   245

Colour*

Following page   78

Landprinting at the coastline: feeding patterns of small crustaceans.
Men . . . are able to suckle infants.
Dolerite coastline at West Cape Howe.
The Snowy River at New Guinea Bend in Victoria.
Below the Snowy (or Jimenbuen) Falls.
Landprinting in south-east Australia: an aerial view of Melbourne.
The main street in Yackandandah in north-eastern Victoria.
Sandhills near Albany, WA, patterned by wind and water.
A ghost gum (*Eucalyptus papuana*) in the MacDonnell Ranges in
     Central Australia.
The gumnut babies, rear view.
A pool in a narrow gorge in the Bungle Bungle National Park in north-western
     Australia.

Following page   174

The passion for neatness; a typical suburban front yard in Perth.
Our brick garden in Fremantle.
The Royal Crescent at Bath.
Melaleuca lanceolata creates a version of the Arcadian at Phillip Island, Victoria.
A National Parks Ranger in Victoria with rabbits, cat and fox.
The Laidlaw Range in the Canning Basin, WA.
A 'dragon' or agamid lizard (*Pogona vitticeps*), near Alice Springs.
Termite mounds in south Kakadu National Park.
Fire in the mallee.

*These plates are available for download in colour from
www.cambridge.org/9780521659994

Foreword

by G. J. V. Nossal

C. P. Snow's two-culture gap keeps popping up in all sorts of new ways. The chasm narrowed substantially when dramatic events such as the taming of DNA and the microelectronic revolution caused intellectuals within the humanities to focus more on science and technology. But fresh cracks originate from epicentres of incomprehension (structuralism, for example, or the arcane jargon of the new biology); they spread and threaten. If we value true learning, if we wish to continue humanity's age-old, unrequitable quest for seamless knowledge and understanding, we need fissure-menders, voices which challenge but unify.

George Seddon may well be Australia's prime example of this rare species. Educated initially in English, he was sufficiently fascinated by science to obtain a doctorate in Geology. His university career has seen him within departments as diverse as English, Philosophy, Geology, History and Philosophy of Science, and Environmental Studies. His practical work has been in landscape planning, urban design and even oil exploration. He is sought all over the world as a keynote lecturer on many topics. Throughout the last quarter-century or more, he has been a prolific writer. *Landprints* is a synthesis of this scholarship. The book resists classification, for it is much more than a collection of essays. Specialised technologies lessen the scope for general debate. 'As the tools of analysis have become sharper, the range of discourse has shrunk, and has tended to become one of professional set pieces.' So Seddon sets out to widen discourse, and the array of diverse past works is grouped, linked, revised and enlivened by introductions and conclusions. Despite the 'linearity of language', Seddon achieves a 'polyphonic account', a fugue for six voices, as his eclectic yet disciplined reading illuminates his search for understanding of language and landscape.

What is *Landprints*? It is history; it is artistic and literary criticism; it is natural history and philosophy. It reveals many principles of landscape architecture, conservation, geology and ecology while avoiding the dryness of most textbooks. It is above all a search for general perspectives from particular examples. It is a valiant attempt to confront some of the key paradoxes of the day: internationalism, liberating but homogenising; regionalism, capable of cementing a sense of place, but possibly parochial and restricting – 'McDonald's setting up shop in China on the one hand, the horrors of ethnic cleansing in Bosnia on the other'. Seddon eschews simplistic solutions. Loyalty and commitment, regional culture, neighbours and neighbourhood must somehow be made to co-exist with academic culture, seen as international, 'the enemy of the parochial; its habitual mode is ironic detachment'. As regards the landscape, both Arcadian and Utopian visions represent oversimplifications. Seddon recognises the place of the human in the landscape, revels in the interplay between natural forms and human creations. The sweep of a freeway, a Robin Boyd circular building faithfully reflecting 'bare rounded hill forms behind and around it', a church spire with the backdrop of the Blue Mountains, even human figures in a landscape painting can enhance rather than destroy the harmony of the scene. Indeed, the same rainforest can seem luminous and paradisal to Jeannie Baker but penitential, dark and menacing to Judith Wright. We must understand such ambiguities and indeed the complexities of deep human attitudes if we are to approach conservation issues intelligently.

'We should be searching first for clarity', and this Seddon does consistently, chiding us for our ready espousal of clichés and oversimplifications. Our landscape is 'vast', 'harsh', 'hostile' and 'unforgiving'. Vast in comparison with a Europe which embraces Russia? Harsh, hostile and unforgiving for the Aborigines? Seddon's clarity in and respect for language is punctuated by some stunningly original – indeed improbable – images. Thus: 'Can you imagine the Parthenon getting planning approval today?'; 'we are not only "future eaters"; we also devour the past, a kind of tourist destination'; or 'three persistent dreams of place; the first, of the Mediterranean, the second, of coral islands in the Pacific, and the third, of tropical jungle'. The rightness of each thought is immediately illustrated by examples, usually with a quite extraordinary juxtaposition of ideas and not a little whimsy.

Seddon displays a courageous balance in his promotion of environmental concern. Already in 1972 he saw the flaws of extremism, such as the bleak predictions of the Club of Rome. Nevertheless, he avows that it is silly to divorce emotion from the environmental debate. What's wrong with being passionate about an important subject? But emotion should co-exist with reason. Above all, we must avoid 'contempt for and alienation from Western society'. My favourite sentence in the book: 'It is counterproductive to destroy the faith of the young in their own society'. How often do we hear this from a conservationist?

George Seddon is, in the end, passionately and quintessentially Australian. 'Australia is inside our heads, your Australia in yours, my Australia in mine'.

The analysis of Australia's ecological fragility, the description of continuing threats to flora and fauna, avoid the shrillness of many authors and somehow gain strength from that. The lampooning of Australian attitudes and stereotypes is softened by a youthful enthusiasm, a zestful optimism which illuminates the whole work.

As I pen these lines, Australia is facing a wide-ranging inquiry into tertiary education. Among other things, the inquiry will seek to define the true purpose of universities, the distinction and balance between education and training. I wish every member of that panel could read *Landprints*. I will end with a line which George Seddon applied to John Passmore's *Man's Responsibility for Nature* in 1974. The publication of *Landprints* is a major intellectual event.

Acknowledgments

When the essays in this book have been previously published, acknowledgment is given in a footnote on the first page of the chapter.

Photographic credits are given *in situ*. Here I would like not only to thank but to celebrate four friends for their outstanding image-making capabilities: Colin Totterdell, Richard Woldendorp, Michal Lewi and John Hanrahan. I am deeply grateful to the Trustees of the Norman Wettenhall Foundation for a generous subsidy towards the cost of colour reproduction, which would not otherwise have been possible. I am also grateful to Julia Ciccarone and the Robert Lindsay Gallery, Melbourne, for their generous permission to reproduce the image of 'Birth of the Australian'.

One of the privileges of the academic life is the stimulus of ideas from friends and colleagues, not, of course, all themselves academics, but without whom the life of the mind would be hard indeed. So I thank many unnamed people and name two: Tom Griffiths, who encouraged me to think (on and off) that such a collection might be worth assembling; and Basil Balme, who read it in draft. I owe much to Cambridge University Press, especially to Phillipa McGuinness and Jane Farago, and to Sally Nicholls and Ron Hampton. They have given me a hitherto undreamed of sense of the cultural centrality of Oakleigh.

Permission to quote is gratefully acknowledged from: Paul Carter, *The Road to Botany Bay*, Faber & Faber, 1987; Peter Carey, *Oscar and Lucinda*, University of Queensland Press, 1988; Tim Flannery, *The Future Eaters*, Reed Books, 1994; Michel Foucault, *The Order of Things*, Tavistock Publications, 1970; Tom Griffiths, *Hunters and Collectors*, Cambridge University Press, 1996; John Scott, *Landscapes of Western Australia*, Aeolian Press, 1986; Glen Tomasetti, *Thoroughly Decent People*, McPhee Gribble, 1976; Maureen Walsh, *May Gibbs: Mother of the Gumnuts*, Angus & Robertson, 1985; Don Watson, *Caledonia Australis*, Collins, 1984.

For Joan and Norman Wettenhall

Prelude:
Dual allegiances

A Frenchman looks at 'les Australiens'

In the early seventeenth century, a Frenchman, Jacques Sadeur, spent 35 years in Australia (or rather, *la terre Australe*). When he eventually got back to Europe, he wrote a book about it, first published in Geneva in 1676.

He describes our national character rather well – for example, one Australian tells him that: 'we make a profession of being all equal, our glory consists in being all alike; all the difference that there is, is only in divers exercises to which we apply ourselves' (presumably he is thinking of rugby versus soccer versus Australian Rules football).

The book contains '*les coutumes et les moeurs des Australiens*' – our customs, our studies, the special animals of the country and so on. There are only four large animals, and none of them dangerous, there are no venomous serpents, there are no troublesome insects, and in particular, specially mentioned, there are no flies. There is a magnificent range of mountains along the south coast, much higher than the Pyrenees, and some broad rivers. It snows occasionally in the south, but not much, and people go around without clothes all the year, since it is never excessively hot nor excessively cold. We live on fruit that ripens the year round. Our houses have exquisite floors and walls of translucent stone (like alabaster, but crystalline-hard and beautifully veined). We are brave warriors, but generally peaceful. Everyone seems to love everyone else, but no one person in particular, and it is not clear how children come into being, since it is a crime to discuss the procreative act. 'Men' as well as 'women' are able to suckle infants, so gender roles are blurred (see colour illustration following p. 78). And the countryside everywhere is a garden.

Things have changed a bit since the early seventeenth century. Needless to say, both the story and its narrator are an invention, an imaginary voyage in the tradition of Swift and many others, and, like most of the imaginary voyages, it is also a Utopian satire on the contemporary world. The Australians are a much better lot than the corrupt and violent society the author saw around him in seventeenth-century Europe. He does not make his Australians quite perfect, however: one way to achieve a satirical contrast is simply to make them very different in some supposedly basic aspect of human behaviour so that we reflect that what we take to be basic may be little more than arbitrary custom. Samuel Butler does this two centuries later in *Erewhon*, set in an imaginary New Zealand, where criminals are sent to hospitals, and the sick to prison.

The real author is not Jacques Sadeur, but Gabriel de Foigny, who was born in 1630 in a small village in the Ardennes. He received a good education, entered a monastery and became a preacher, but was soon unfrocked for unfrocking too freely among the girls of the neighbouring village. France became too hot for him, so in 1666 he went to Geneva, forswore the Catholic faith, and converted to the puritanical creed of Calvin. He soon got into trouble again, adding drunkenness to licentious behaviour. His book was also to get him into trouble in Switzerland, and eventually he went back to France, where he died in 1692.

I had come across de Foigny's book from occasional quotations and references in other works, but I had never read it through until I became aware of an exhibition in Melbourne by a young Australian-Italian painter, Julia Ciccarone, who has illustrated Sadeur's text; her work, and the excellent catalogue that goes with it, are based on an English translation published in 1693 (with the fictitious author undergoing a name-change from Jacques to James). This translation is based on a second edition in French, published in Paris in 1692, somewhat expurgated from the 1676 edition that so shocked the good folk of Geneva.

Some of de Foigny's detail is standard Utopian, some highly original. We Australians are born both good and free, so we need a formal government no more than we need a formal religion. We meet to discuss the affairs of the community, but we have no written laws and no rulers. There is no private property, and the family is no threat to the unity of the community, because it does not exist. These Utopian ideals were tried by a group of Australians in the last century in Paraguay, and by the Israelis on their kibbutzim.

The sexual behaviour of the Australians is one of the highly original features, for we are all hermaphrodites. Sexual relations except in the service of reproduction are regarded with horror and reproduction itself is something of a mystery, not to be discussed, although every being is expected to produce one child. Presumably these attitudes to sex are a parody of the extreme puritanism of Calvinist Geneva. In other respects, however, the Australians are very open about their bodies. Everyone goes around naked. Sadeur discusses this with an Australian, pointing out that Europeans are quite shocked to see a person naked. He alleges that the reasons for the European attitudes are 'modesty, the rigour of the season and custom'. Sadeur told the Australian:

> there were some countries amongst the Europeans, where the cold was so insup-portable to the body, which was more delicate than that of the Australians, and that there were some that even died upon it, and that it was impossible to subsist without clothes: I told him that the weakness of the nature of either sex was such that there was no looking upon one that was naked without blushing and shame, and without being sensible of such emotions as modesty obliged me to pass over in silence. (de Foigny, 1693, p. 74)

But his Australian interlocutor was not impressed by these arguments:

> Is not this to father upon all the world what is contrary to Nature? We are born naked, and we can't be covered without believing that it is shameful to be seen as we are: but as to what thou sayest concerning the rigour of the season, I can't . . . give any credit to it; for if this country is so insupportable, what is it that obliges him that knows what reason is, to make it his country? (de Foigny, 1693, p. 75)

This seems to catch the authentic Australian voice as heard today in England or France: 'if the climate is that bad, why stay there?'

There is a lack of differentiation between the parts of de Foigny's Australia that reads like a good guess:

> What is more surprising in the Australian Dominion, is that [apart from the great fringing range along the south coast, from which the rivers run, guided by brilliant projects of irrigation engineering] there is not one mountain to be seen; the natives having levelled them all . . . this great country is flat, without forests, marshes or deserts, and equally inhabited throughout . . . To this prodigy may be added the admirable uniformity of languages, customs, buildings and other things which are to be met with in this Country. 'Tis sufficient to know one quarter, to make a certain judgement of all the rest; all which without doubt proceeds from the nature of the people, who are all born with an inclination of willing nothing contrary to one another; and if it should happen that any one of them had anything that was not common, it would be impossible for him to make use of it. (de Foigny, 1693, pp. 51–2)

Perhaps we are a little on the conformist side, but de Foigny still likes us, and makes an extended comment on the superiority of our manners and customs over the rest of the world. Even if he sounds like a Channel 7 reporter on the Australian performance at the Olympic Games in Atlanta, he is worth hearing.

> I could not but admire a conduct so opposite to our defective one, that I was ashamed to remember how far we were from the perfection of these People.
>
> Our best morality is not capable of better reasoning, nor more exactness, than what they practise naturally without rules . . . The insatiable thirst after riches, these continual dissensions, these black treasons, bloody conspiracies, and cruel butcheries, which we are continually exercising towards one another; don't these things force us to acknowledge that we are guided by passion rather than reason? Is it not to be wished that in this estate, one of these men which we may call Barbarians would come to disabuse us, and appear in so much virtue as they practise purely by their Natural Light, to confound the vanity which we draw from our pretended knowledge, and by the assistance of which we only live like beasts. (de Foigny, 1693, p. 78)

This is a Utopian dream, but we have, at least in my view, realised a small part of it. Australia at the end of the twentieth century is less violent, less cruel, less tyrannical, less intolerant than France at the end of the seventeenth, although perfection still eludes our grasp, Channel 7 notwithstanding.

Obviously the Utopian fictions tell us nothing about the real Australia in the seventeenth century. What they do tell us is that, in Robert Hughes' phrase, Terra Australis was Europe's 'geographic unconscious', and in that they may tell us quite a lot about the real Australia at the end of the twentieth century, since we are still reinventing ourselves as a continent and a people. That reinventing relates partly to the physical reality of the place, and partly to the expectations, hopes and dreams our forebears brought with them. We redefine ourselves both positively and negatively against a primarily European past.

The European past and present is our cultural and intellectual heritage, not as a kind of finishing school, but as the key to an adequate understanding of many current dilemmas. To understand the force of reductionism in science and some social and economic thinking, you must go back to Descartes; to understand the habits of resource consumption of the Western world, you must study, among other things, the history of the Pleistocene in Europe, and read *Capital and Material Life 1400–1800* (Braudel, 1974); to understand our responses to landscape, study the history of Western art and read Jane Austen.

Our European patrimony is also enriching: to think about the possibilities of urban design, it is well to know Palladio's city of Vicenza, and Aix en Provence,

Seville and Bath – and then there is Venice. Our lives would be unimaginably bereft without Bach and Mozart, Shakespeare and Tolstoy. All of this imaginative world, however, creates images of a human habitat that are sharply dissonant with our own. Here lies the rub, the inescapable tension in being or becoming Australian.

The tension is not peculiar to us. It is one of the continuing themes of Henry James' novels, in which his characters are torn apart by a dual allegiance. Yet it is even harder for us. Despite the differences between European and North American society, the two lands have many physical and biological links, far more than is the case with Australia and Europe, which do not even share something as primal as the seasons.

Dual allegiance is also the theme of Henry Handel Richardson's novel, *The Fortunes of Richard Mahony*. When Mahony, her protagonist, has had enough of the coastal fog and social hierarchy of Buddlecombe in the south of England, he longs nostalgically for 'a really sweet apple' – such as is to be had in Australia – instead of a 'specimen that's red on one side only. I believe England will stick in my mind, for the rest of my days, as the land where the fruit doesn't ripen' (p. 418). The reader remembers, as Mahony does not, that he had longed to sink his teeth into a real English apple some two hundred pages and five years earlier in Ballarat. The reader also knows by now that, whether distant apples be greener or sweeter, it will take more than apples to make Mahony feel at home in either hemisphere. He packs his discontents with him, some of them divine, some mundane. Thus the tag from Horace: 'Those who run across the seas change only the heavens and not their minds', which is recollected by Mahony (in Latin: *Coelum, non animum, mutant, qui trans mare currunt*) when he makes his decision to return to England (p. 319). The phrase means one thing to the reader, and another to Mahony, who uses it to justify the thought that he should never have left England in the first place, whereas we know that it will be as true of the return voyage as it was of the voyage out.

It cannot, however, be wholly true, in the sense that the mind or spirit is unchanged by circumstance. Mahony himself has been changed by Australia, and it is this that in part unfits him for a return to the closed society of rural England. Perhaps we could say that his Australian experience determines the particularity of his discontent, but the discontent itself, the result of individual and cultural background, remains constant. Much of this remarkable novel is an exploration of the range of applications of the Latin tag – of what changes and what is constant as we move between the two hemispheres. That is one of the themes that runs through many of the essays in this collection, with the complementary concern to work out what is enabling, and what disabling, in our inherited cultural traditions, especially in response to conservation and landscape planning issues.

Conservation battles, conservation debates

World War III is upon us. What is at stake, we hear, is the very survival of our species. Speaking as a geologist, I wouldn't bet on it, but we could win a reprieve for a few more millennia. I would be asking for a great deal more than survival, however. I want a rich and varied life for our species in a rich and varied planet, sharing it with a host of other species.

As with most wars, the objective is clear – we want to win – but the goals are not. What do we want to win, from whom, and above all how and at what cost? The trumpets may sound the opening bars of the chorus of that most stirring of all songs, the *Marseillaise*: '*Aux armes, citoyens*', but we remember the Terror, and all the other costs of that great cry to achieve great ends. In this war, the enemy is ourselves, a complication. All wars are confusing, and they are always accompanied by a war of words and of images which aim to reduce or eliminate complexity. This is partly why their costs are so immense. Think how convenient and manipulative is a word like 'enemy'. Once the enemy is identified – in our case, the Germans, the Italians, the Japanese, then the North Koreans, the Viet Cong, the Iraqis – you can bomb Dresden, Hiroshima, Baghdad, deforest much of Vietnam. One simple word conceals the reality of incinerated men, women and children.

The essays in this book are about landscape, usually Australian landscape and its interpretation. The potter may leave her thumbprint on the base of her pot as an identifying mark. A human footprint in the sand was electrifying news to Robinson Crusoe, and its immediate meaning was unequivocal – he was not alone. But the implications of that immediate meaning were uncertain. None of us is alone, and the sands have been trodden, overprinted, by countless feet. 'Landprints' are the marks on the surface of the earth, many of them made by our species (see colour illustration following p. 78). The sum of the distinctive characters in a given area, whether 'natural' or man-made, can be called a 'landscape'. But although they have a physical substrate, landscapes are also a cultural construct. The ways in which we read them, talk about them, perceive them, work them over, use them, evaluate them functionally, aesthetically, morally: these are all informed by our culture. These are also among the themes explored in these essays.

They are very personal themes. All my life I have been driven by two passions: a love of language and a fascination with the physical world of rocks, trees, rivers, mountains, landform. These two interests have led me into academic departments with different names. My interest in language has led me twice into Departments of English, and once into a Department of Philosophy. My interest in the land led me to take a doctorate in Geology, to an appointment in Geology (at the University of Oregon), and later as the first Director of the Centre for Environmental Studies at the University of Melbourne. Both interests found legitimate expression when I held the Chair of History and Philosophy of Science at the University of New South Wales.

But both have always found expression. My two enduring passions – let us now call them English and Geology – can be seen as mutually enhancing intellectual modes as much as scholarly disciplines. Geology first. It insists on *la longue durée*, the long-term perspective on events, which is why Fernand Braudel is my favourite historian. The last ten thousand years are formally defined in Geology as Recent. In some senses, Geology is ineluctably pragmatic. Mineral deposits, for example, are where you find them, not where you would like to find them or where you think they should be. Geology is driven by meticulous observation in the field, and, in that sense, fact takes precedence over theory; although fact is always tied to interpretation, the inconvenient observation cannot be suppressed, and particularity reigns. Geology proceeds by multiple working hypotheses, rather like the diagnostic procedures of the experienced and skilled physician. The more possible explanations you can think of, the better. You can then proceed to knock out a few, but may still be left with complementary 'causes' and some uncertainty. There is an irreducible complexity about the natural world.

Geology is global. Political boundaries are irrelevant. The Precambrian in Labrador, Scandinavia and Western Australia is all Precambrian. Granite is granite, wherever you find it. Finally, Geology inclines to a measure of determinism, another bias towards pragmatism, in that not everything is possible. The geology of an area in combination with the climate determines the soils, and the soils, the vegetation; between them, they profoundly influence human affairs. Most historians describe the play and the players, but ignore the stage. Not Braudel, nor Blainey, nor the economic historians.

English as an intellectual mode shares some of these characteristics, especially the taste for particularity, the delight in the infinitely varied living earth. It is also profoundly humanistic, endlessly fascinated by human beings, including all those faces that are usually kept hidden in polite society. As an academic discipline it is, in a way, subversive of discipline. Beneath the robes of the Nobel laureate in Physics, the judge, the bishop, the murderer, the butcher, the vice-chancellor, the dustman, there is always that 'poor forked thing, a man'. The search for that kind of truth, the truth of the experience of living, ignores the signs that say 'Keep Off the Grass'.

Applied conservation

A good part of my working life, however, has been spent in applied conservation and landscape planning for mundane projects such as identifying satisfactory routes for power lines (I took part in route selection for nearly every 500 kV power line in Victoria for two decades, with involvement also in Western Australia, South Australia, Queensland and the Northern Territory: by force of circumstance, I became the 'power line man' in landscape planning). Other projects included urban design guidelines for Hawthorn, landscape planning for the southern Mornington Peninsula, strategic planning for Phillip Island (all these in Victoria), a review of public open space planning for Canberra, and so on. These were all commissioned reports and therefore had clients, who were themselves answerable to a broader public. There were many such reports and studies. Looking back, I say with satisfaction that my – or our, since it was the Centre for Environmental Studies at the University of Melbourne that was commissioned and we worked as a team – recommendations were nearly all accepted and put into place.

The reports show a controlled idealism: what I tried to do is to understand the genuine constraints, gauge the intentions of the clients, and pitch the recommendations two notches higher than they had thought they were prepared to go, but not higher to the point of rejection, taking care to present our case as coherently and persuasively as possible, using good graphics as well as words.

This is all pragmatic. I was content if the results were significantly better than they would have been if we had not been involved. Although local in scale, however, most of the projects also had implications touching on, although not directly confronting, many of the major conservation issues of the day in Australia. These I take to be: soil conservation above all, and water management, catchment management and the state of the rivers (generally in crisis); coastal planning (still more an aspiration than a reality in Australia); and urban planning and design (a visiting

Italian colleague once asked me 'Why don't you build your cities as if you intend them to last for a thousand years rather than as temporary encampments?').

In my private life I have sometimes manned the conservation barricades, and would like to have stood in the way of the bulldozers in the Daintree. I have also tried to modify attitudes through books and essays. The essays and reviews collected here are intended not so much as a direct contribution to conservation issues as a contribution to the conservation debate and a better understanding of landscape planning. I believe in civilised debate: I want intelligent debate, and I want also to live in a civil society, like John Passmore (see Chapter 21, The perfectibility of Nature). It is an immensely complex debate, because conservation and planning issues reach deep into our economy, our ecology, our society, our language. 'The truth', said Oscar Wilde, 'is seldom pure, and never simple'. The usual outcome of simplifying issues is simplistic conclusions, which often have consequences worse than the problem the measure was intended to address.

I have seen this happen again and again, especially in aid programs in Southeast Asia and Papua New Guinea, where I had a review role for a number of Australian Development Assistance Board (ADAB) programs, and also Unesco programs through Man and the Biosphere (MAB). It is extraordinarily difficult to devise and deliver good aid programs to a culture different from ours without disrupting the social fabric of that culture. To give details would need another book. This one is primarily about Australia, but it is difficult here, too, and it is also difficult to display the complexity of the options, since so many responses are pre-programmed.

The survey of the proposed power line from Darwin to Jabiru through Kakadu National Park is a case in point. Ideally, no one, including myself, wants a power line in such an environment, where we go to escape technology for a while, and to submit ourselves to the pulse of more 'natural' rhythms. But everyone, or almost everyone, wants a cold drink when they get there, which requires refrigeration, and the same people want showers, which need electrical pumps (or noisy generators). The National Parks and Wildlife Service needs computers and cold storage, power and light. Many visitors, especially the elderly and overseas guests unused to the enervating climate, want and need air-conditioning – and, of course, Ranger Uranium is there, whether you like it or not, and needs power to run its plant. Ranger provides the power with its own generating station, which also supplies the township and the National Parks and Wildlife Service headquarters. The generating station is diesel operated. The diesel is imported, and government-subsidised – still, in 1997. The consumption is estimated at 650,000 litres annually, to supply one medium-sized power station at Jabiru and twelve small ones, some of which leak, scattered through the region. The diesel is brought in three times a week in huge tankers some two hundred kilometres from Darwin. The Arnhem Highway, the only road, crosses many rivers, by causeway rather than bridges, shallow during 'the Dry', sometimes torrential during 'the Wet', but the tankers must get through. Think of the consequences of just one spill. The electrical power to be carried by the proposed power line, by contrast, is generated from natural gas in Darwin, whence the gas is carried by pipeline from Central Australia, not imported, not subsidised, not significantly polluting.

There were many subsidiary issues other than the two sketched above. The hostile responses to the Draft Environmental Assessment, released publicly, did not as a rule address them: there was an immediate rejection in Melbourne, Sydney and Canberra by some respondents, some of them regarded as the guardians of our environment, to the very idea of a power line in a national park, without any serious attempt to consider the alternatives. Responses such as these do not, in my view, advance the conservation cause, but we are better able to understand them if we see how they are generated by the ambiguities of our attitudes towards nature (this will be explored in Chapter 1, The nature of Nature).

References

Braudel, Fernand, 1974, *Capitalism and Material Life 1400–1800*, Glasgow: William Collins Sons.

Sadeur, Jacques (Foigny, Gabriel de), 1676, La Terre Australe Connue, (Geneve?): Vanne.

Sadeur, Jacques (Foigny, Gabriel de), 1692, *Nouveau voiage de la Terre Australe, contenant les coutumes et les moeurs des Australiens, leur exercises, leur etudes, leur guerres, les animaux particuliers a ce pays, et tout les raretes curieuses qui s'y trouvent*, Paris: Chez Claude Barbin au Palais.

Sadeur, Jacques (Foigny, Gabriel de), 1693, *A New Discovery of Terra Incognita Australis, or the Southern World*, by James Sadeur a Frenchman, London: Charles Hern.

Richardson, Henry Handel, 1930, *The Fortunes of Richard Mahony*, London: Heinemann.

Fugue for six voices

These essays have been grouped under six headings, assigned according to their primary emphasis. The themes, however, are interrelated, weaving in and out of one another, and to show this is an aim of the book. Thus, many essays could have been assigned to a different section.

To take one example, Theme IV, Making: creating gardens and the evolution of styles, is about creating gardens and designing landscapes. But to make a garden or design a landscape, we express our sense of place (Theme III, Locating: the sense of place); our sense of place is a cultural construct, and that has a two-way, feedback loop to the way in which we read the environment (Theme II, Perceiving: the eyes and the mind). Our location and perception are dependent on our experience of the world and on the distinctive character of our habitat (Theme VI, Sharing and caring: ecological frameworks). Cultural constraints are a way of relating to the environment, and affect our behaviour towards it (Theme V, Analysing: ideologies and attitudes). Finally, we communicate our thinking about all of the above in words (Theme I, Talking: the language of landscape).

The words, moreover, are slippery: at times they carry conflicting meanings (see Chapter 1, The nature of Nature). Some of these conflicts are so basic that we cannot eliminate them, and the best we can do is to be aware of them. There is one such conflict in the example I have just given. 'Landscape' is a way of looking at a terrain: it is a perceptual term, not an objective reality. If you doubt this, try asking a farmer and a city bank manager to describe the same paddock – the 'same' landscape will turn out to be very different. The landscapes of this book are my way of looking, hence the autobiographical flavour, so that you have an idea of who is doing the looking. But if landscapes are a way of looking, then we cannot design landscapes. No one else can design how I see. Yet we cannot avoid such phrases for long. We constantly intervene in the 'natural' environment (that is our nature). The ways in which we do so, and the ways in which we read the outcome are mediated by our cultural context, and this leads us to talk about design intentions and design outcomes as 'landscapes'. Language is messy, but it is the best we've got.

The essays assembled here have been written over a span of more than thirty years. They are ordered by theme rather than by the time of writing, but the year is given for those interested in chronology. Whether they show a maturing, I can't tell. On the whole, I doubt it, except in a few special cases. They were written wearing different hats, for a variety of different journals, and thus for different audiences. They represent different phases of my working life: I wrote some pieces as a geologist, others as a historian and philosopher of science, then as an environmental scientist, and, at both ends of my career, as a member of an English department. Yet all through these changes, my main interests were in words, in the physical world and the forces that mould it, in caring for it, and in how it looks – a strong visual response has always been a part of my temperament. So the same themes have been resurfacing, although sometimes with different labels. And that, of course, has stimulated an interest in the labels.

Most of the essays have been lightly edited to link the themes and make the format and style more nearly consistent, although not to the point of eliminating the use of 'man' as a gender-neutral term in historically appropriate contexts. Two essays have been restructured and reduced in length (Chapter 9, Dreaming up a rainforest, and Chapter 13, Cuddlepie and other surrogates). Recent material has sometimes been added in the form of postscripts; about a third of the material is newly written for this book.

THEME 1

Talking: the language of landscape

My mother, or your mother or your grandmother, once said: 'Don't put that penny in your mouth, you don't know who's touched it'. Peter Porter began a talk on language with the same phrase: 'Don't touch that word: you don't know where it's been'. Language carries the riches and the burdens of the past, and the language of landscape, like all language, is loaded. Chapter 1, The nature of Nature, struggles with some of the deeper ambiguities of our language, inescapable because the ambiguities are not merely linguistic: they reflect ancient ambiguities in our cultural attitudes to the world. Chapter 2, Words and weeds, addresses the burden directly.

My essay, 'Imaging the Mind' (Seddon, 1993), while not at first sight about landscape at all, extends the  theme of these two chapters, and might have been titled 'Landscapes of the Mind'. Nothing gives a more interesting glimpse of our world view than the way we image the workings of our own minds. To describe someone who 'is not quite all there' as 'having kangaroos loose in the top paddock' illustrates the centrality of landscape imagery in our culture, contrasting nicely with the English 'bats in the belfry' which conjures up a very different world. Space forbade its inclusion in this book.

Chapter 3, Journeys through a landscape, allows the opportunity to look at some of the clichés about Australia, which dull rather than sharpen perception; it also looks at visual images rather than words as a means of communicating responses to the landscape – but we still come back to words when we try to compare our responses to the images.

Chapter 4, On *The Road to Botany Bay*, is a review of a dense but rewarding book by Paul Carter. It shows how often our accounts of the past falsify the reality of that experience; for example, the school-book accounts of the exploration of inland Australia by Eyre or Leichhardt, Giles or Warburton, show their routes by dotted lines across the map of the continent. But for them there was no map. Each day pushed into a void.

Carter's book is relevant to the themes pursued in this book: for example, he contrasts the naming practices of Cook and Banks. He presents Banks' practice of giving a name to a plant as a way of disposing of it – once named, it is 'known to science', ordered and categorised in the Linnaean system, which is, of course, European in origin. The plant is then owned and no longer foreign. He presents Cook's naming, by contrast, as a form of discovery, recognition of the new, of the richness and diversity of experience. I have reservations about the details of the examples, but not about the contrast they illustrate, or Carter's skill in making the two men carry with them on their long journey an echo of one of the great philosophical debates of their century, that between the realists and the idealists. Thus the cultural currents of their Europe are felt beyond the furthest shores of the known world, and moderate their responses to a new world that lay well outside the experience of that culture.

All the essays in Theme I are fairly recent, the oldest being Chapter 5, A Snowy River reader, which describes the structural and linguistic problems I met in trying to write an environmental history of the Snowy. In time, the history was written, and it was published in 1995 as *Searching for the Snowy: An Environmental History*, in which I tried to get at the river from every angle; yet I added a footnote to the invitation to the launching of the book: 'I still haven't found it, and if anyone has seen it, please let me know. A sketch map might help'. Of all the responses to the river, the most expressive I came across was that of the tough canoeists: behind their sparse, dry, laconic speech looms the actual presence of the river. You can almost hear its muted thunder in the gaps between the words.

References

Carter, Paul, 1987, *The Road to Botany Bay: An Essay in Spatial History*, London: Faber & Faber.

Seddon, G., 1993, 'Imaging the Mind', *Meanjin*, vol. 52, pp. 183–94.

Seddon, G., 1995, *Searching for the Snowy: an Environmental History*, Sydney: Allen & Unwin.

1 The nature of Nature*

> *How do you know but ev'ry Bird that cuts the airy way,*
> *Is an immense World of Delight, clos'd by your senses five?*
> (William Blake, 1793)

The range of meanings of 'Nature'

The problems in understanding how we ascribe meaning to a key word like 'Nature' are deeply embedded in our cultural history – from which, of course, the word derives its complex meanings, as with all words. We might begin with a warm-up exercise familiar to philosophers. What are the antonyms to Nature and the natural? With what is it in contrast? The major pairs are as follows: the natural and the supernatural (or the Divine); the natural and the unnatural; the natural and the human (as in Man and the Biosphere, or Man and the Environment, so that Nature becomes everything that is Not-Man); and the natural and the artificial. Wool is a natural fibre, nylon is an artificial one, although both are man-mediated. The problem is even more obvious when we think of the advertising campaign claiming that sugar is a natural food, and all those television advertisements showing waving fields of golden sunflowers, or whatever, about to be processed into margarine. Logically, butter and margarine are equally synthetic or equally natural, but the difference in our feelings about the two shows clearly that we are already beyond the bounds of logic.

The first of the 'meaning pairs' (the Natural–the Supernatural) has first place in the history of Western thought and is still very powerful, but it is no longer the primary contrast. The view that Nature was inferior to Super Nature, the Supernatural, the Divine world, co-existed with its opposite, that Nature itself is Divine, through most of the Middle Ages and beyond, while the view that Nature is an expression of Divine creative power has persisted to the present day. Contempt for the natural world – the *contemptus mundi* – was exemplified by the lives of the saints; in its extreme form, it denied value to the natural world, to the self, and to all pleasure, especially sexual pleasure.

*First published as 'The Nature of Nature', *Westerly,* no. 4, 1991, pp. 7–14.

The Divinity of Nature

However, the *contemptus mundi* was never uncontested. Since God was Creator, and the Perfect Being, then his creation might be seen as perfect, and the natural world, it might be argued, as fit, not for contemptuous dismissal, but for study and delight. These opposites were to an extent reconciled by the concept of the 'Great Chain of Being'. 'Everything, or nearly everything, that existed was thought of as necessarily existing, but as graded in value. The further away from God, the lower the value' (Brewer, 1972, p. 28). Nature constituted an artistic order. Sir Thomas Browne put it splendidly in 1642: '*Natura nihil agit frustra* [Nature does nothing in vain] is the only indisputable axiom in philosophy; there are no grotesques in nature, nor any thing framed to fill up empty cantons and unnecessary spaces' (Browne, 1972, p. 16). His reasons, however, are not ours: Nature does nothing in vain because it fulfils God's purposes, which are wholly focused on Man. God made the ants to teach us industry and thrift, the bees to teach us the principles of social order. That is what they are for.

That Nature partakes of the Divine is a component of many non-Judeo-Christian theologies, especially in Asia, and of the animistic world view of Australian Aborigines and other groups, and to the extent that we see God revealing himself through Nature, it may also be a component of Christian theology.

These are deep waters, but before we strike out for the shore we should note that the romanticisation of Nature as partaking of Divinity is a major component of popular culture today – it is very common. It is expressed in phrases such as 'interfering with Nature' (which is supposed to be a bad thing to do) or 'Design with Nature' (which is supposed to be a good thing to do); and, of course, in the more basic and long-lived phrases such as 'Mother Nature' and 'Nature knows best', beloved of the homoeopaths. The trouble with all these phrases is not that they are wrong, but that they are fuzzy-minded. They all express a grain of wisdom, but a wisdom that is applicable in some situations, and not in others. As with all slogans, they do not carry with them any instructions that indicate when they should be used – criteria of application and misapplication. Consider 'Nature knows best' (presumably because Mother knows best). As a general warning against interventionist medicos who have an inbuilt tendency to overtreat, this is wise. As advice to parents of a child with acute appendicitis, it is dangerous folly. Or consider 'Design with Nature'. As general advice to a society that has so often turned to engineering and technological solutions to biological and social problems, this is again wise advice. But there are just as many circumstances in which Nature is inimical to our purposes, and we therefore design against her (if one cares for this turn of phrase, which I don't).

Another version of Nature as partaking of the Divine was a component of the Romantic Revival. Wordsworth is full of it, although his version of Nature takes meaning largely from his detestation – in most moods – of London, 'The Great Wen', as Cobbett called it, a cancerous growth. For Wordsworth, Nature included agricultural landscapes, country folk and children. He did not draw our current distinction between 'natural' and 'cultural' landscapes: his primary distinction was between the natural and the urban. In the United States, Walt Whitman shared somewhat similar sentiments, but the Divinity of Nature is perhaps most fully expressed by the National Parks movement and the language of John Muir and some early members of the Sierra Club. The mood is caught by the superb photographs of J. B. Jackson and, later, Ansell Adams. As David Lowenthal, one of our most subtle cultural-historical geographers has pointed out, Americans were acutely conscious in the nineteenth century that their continent lacked the great cathedrals and other architectural treasures of Europe, so they sanctified their natural monuments instead: the Grand Canyon, Old Faithful, Muir Woods and Yosemite were older and grander expressions of the sublime than anything Europe could show. We have constructed an 'Ayers Rock' cult in the same vein.

Nature as the Enemy

The view that Nature is Divine or Holy co-existed with and was in part reaction to its opposite: the view that Nature is the Enemy. This has always been a part of the popular culture, with good reason, because ordinary people have always been the most vulnerable to the vagaries of natural forces. It has had strong expression in the high culture in periodic mode, like Halley's Comet. The two moods co-exist, but one is now in the ascendant, then the other, depending on a whole range of associated shifts in cultural mood (see Blainey, 1988). Nature as the Enemy is expressed in phrases like 'Taming Nature' (of which we have done a good deal in Australia), or 'harnessing' a natural resource, such as the Snowy River, or the wind, or the tides, which are like a wild horse before we introduce the bit. The popular culture is rich in such phrases, and the gardening columns are full of them: 'untidy' trees for example, or trees that have a 'poor habit', usually said of poor old *Eucalyptus macrocarpa*, and pruning to retain 'a good form' and so on. All of these suggest that Nature at the very least is undisciplined, and much in need of our control.

9

This view lies deep, and can lead to striking inconsistencies. I discovered one in myself recently. I regard myself as ecologically enlightened, but I still don't like to see dead wood on trees in my garden, and I hate to see my *Eucalyptus erythrocorys* disfigured with lerp, small sucking insects of the family Psyllidae. Yet I love birds. A young ecologist a few weeks ago pointed out that lerp are good: their sugary secretions bring insects and the insects bring the birds. So I am learning to love lerp. After all, it's 'natural'.

The natural and the unnatural

However, the antonym pairs that seem to me to underwrite most current discourse using the word 'Nature' are natural–unnatural and 'the natural' and 'the human', in which Nature is Not-Man. The idea of the 'unnatural' has been around for a long time, and although it grades into the 'Nature knows best' nexus of meanings, it usually carries specifically moral overtones, as in 'Sodomy is an unnatural practice'. That not many people use that expression today does not mean that the concept of the 'unnatural' has disappeared, but rather that its range of application has changed. Some will now say, for example, that celibacy is unnatural, or that it is not natural for a young girl to lock herself away in her room reading books all day, or whatever. The point, of course, is that what we consider to be 'natural' and 'unnatural' changes through time. Our concept of Nature is a cultural product.

Nature as the non-human world

This leaves me with what is today the most elementary meaning of Nature: the non-human world – it is in this sense that we talk of the conservation of nature, and understanding natural systems (a phrase I have used myself already). We could hardly communicate without some such distinction, since we could not talk about an undifferentiated cosmos, and one of the most basic distinctions is between the Us and the Not-Us. Yet there are some major problems with this distinction.

It may be surprising to learn that this sense of the words 'Nature' and 'the natural world' is of fairly recent origin. Michel Foucault claims that the landmark is the publication in 1657 by Jonston of a *Natural History of Quadrupeds*. He uses this date to mark the birth of natural history. Before that date, there were just histories; for example, a *History of Serpents and Dragons* by Aldrovandi, or *An Admirable History of Plants* by Muret. Up to and including Aldrovandi:

> History was the inextricable and completely unitary fabric of all that was visible of things and of the signs that had been discovered or lodged in them: to write the history of a plant or an animal was as much a matter of describing its elements or organs as of describing the resemblances that could be found in it, the virtues that it was thought to possess, the legends and stories with which it had been involved, its place in heraldry, the medicaments that were concocted from its substance, the foods it provided, what the ancients recorded of it, and what travellers might have said of it. (Foucault, 1970, p. 129)

The distinction that we make so easily between the knowledge derived from direct observation, that from reliable secondary sources, and that from sources that we regard as legendary or fabulous did not exist. Thus the essential difference between Jonston and Aldrovandi is not that Jonston knew more. In a sense he knew less. The difference lies in what he left out.

> The whole of animal semantics has disappeared, like a dead and useless limb. The words that had been interwoven in the very being of the beast have been unravelled and removed: and the living being, in its anatomy, its form, its habits, its birth and death, appears as though stripped naked. (Foucault, 1970, p. 129)

There has been a further change within our own times. Most of us now wish to see ourselves as a part of Nature, and this new sense of the interdependence of all living systems and their further dependence on physical cycles is a significant intellectual advance – but of course it undercuts the dualism of Man and the Biosphere or Man and Nature! Those signs on freeways, fairly common in Australia, that read 'Animals prohibited on this freeway' now seem comic, although we know that 'animals' means 'horses', and excludes ourselves.

Another aspect of this distinction is one of the enduring puzzles of philosophy, usually approached in the philosophy schools through an introduction to Locke, Berkeley and Hume. Locke was a champion of the newly emerging scientific methods, based on observation and measurement. He was called an empiricist and, later, a realist, in that he believed that there is a real world out there, which we can learn about scientifically, while all the learned, wordy, theoretical debates were a waste of time: we should burn the books. But Bishop Berkeley, labelled an 'idealist', asked an unanswerable question: How do we *know* that there is a real world out there? All we have are our perceptions. We can never know what corresponds to them 'out there'. There may be no 'out there'. Thus the distinction between Us and Not-Us is fallacious. Dr Johnson asserted the reality of the external world by kicking a table, thus splendidly missing the point, but nevertheless reaffirming the common sense of the ages. William Blake asks the Berkeleian question, in poetic form, with the epigraph with which I began (just as Plato had asked it long before either).

> How do you know but ev'ry Bird that cuts the airy way,
> Is an immense World of Delight, clos'd by your senses five?

The answer, of course, is that you don't, and can't, although as I watch a willy-wagtail on my lawn, I wonder. To me, he seems to be *playing* and enjoying it immensely. He is also catching insects by setting them up with his rapid movements, and perhaps courting an unseen mate. You can say that I am giving my perception an unwarranted reality in saying that he is 'there' at all, and that I am projecting my own feelings on to his behaviour to say also that he is playing, enjoying himself. But you can't prove that he is not there or not playing, and I can't prove that he is.

The battle between the idealist and the realist philosophers has not been settled, but the commonsense view that there is a world external to ourselves continues to be held by ordinary mortals. Just what that world might be, however, has become increasingly uncertain over the years. We now know, for instance, that very few animals share our binocular colour discriminating vision; many organisms quite literally see a different world.

Even our most basic perceptions, such as the judging of distances and relative size, which we take to be objective responses to the real world, are in fact conditioned by our prior experience. In a new environment, our prior experience may turn out to be irrelevant or even substantially misleading as, for example, experience in England led early settlers in Australia to 'read' tall trees as an indication of fertile soil. Generally, we see what we need to see or want to see. We have to learn to see in an unfamiliar setting; seeing is not like taking photographs. I still remember clearly how I learned to see zebra in tall grass in Africa. I was trekking on foot with Ian Player, the conservation conscience of South Africa, in the Black Umfolozi in northern Natal. I was told, or whispered to (for talking was out; there were lions and black rhino around) that we were near a herd of zebra. At first I couldn't see a thing. Then I saw a little flick, which was a twitch of the tail, and then the zebra came into focus. The South Africans I was with just saw zebra; if there was no consciousness of the two stages in the process of recognition for them, it was because those stages were so speeded up as to seem instantaneous.

Language structures our map of reality

Thus individuals perceive different worlds, and whole cultures do so on a greater scale. Whether or not we are realists, philosophically, we are driven towards relativism by discoveries in science, including physics and psychology, and increasingly also by linguistic theory. The old chestnut about the Eskimos having a dozen different words for 'snow' has been around for years, but there are many other examples. Aborigines from south-western Australia had about eight different words for 'burning-off' – burning the bush – since they did it in different ways, at different times, for different purposes. In short, language structures our map of reality. The new relativism pervades contemporary culture. Semiotics, the theory of signs, for example: signs are susceptible to a range of interpretation. In much literary theory, we find 'the text', and 'readings': there is a multiplicity of readings, and although some may be of wider interest than others, we are now wary of using words like 'misinterpretation'. A reading is simply a reading.

Scientists are not immune from such analyses. They strive for what we (usefully) call objectivity by focusing on the measurable and the repeatable and this can achieve a degree of paradigmatic consensus, but they too are a product of cultural conditioning. This is a subject of interest to me, because much natural science shows a bias towards the northern hemisphere, which I call Eurocentrism, discussed in Chapter 7, Eurocentrism and Australian science. One illustration will suffice here. There are some six hundred species of one single genus, *Eucalyptus*, whereas there are only six representatives in Australia of one whole family, the Rosaceae, insignificant plants like the bidgee-widgee, *Acmena anserinifolia*. Had the Linnaean system of binominal classification been developed in this hemisphere, there can be no doubt that the families would be different. Part of the problem is that North America belongs essentially to the same biogeographic province as Eurasia, to which it was linked by land during part of the Pleistocene, as the fauna and flora show (oaks and elms, wolves and bears, etc.), and this has tended to confirm the Eurocentric bias of the natural sciences. The point is beautifully illustrated in a short article by Stephen Jay Gould from Harvard, with the title 'What's Wrong with Marsupials?'. The answer, of course, is that there is nothing whatsoever wrong with marsupials, except that they have long been regarded as second-rate citizens. From the outset, they were looked on as freaks; the platypus was actually called a *'lusus naturae'*, and the first specimen sighted in Europe was regarded with suspicion, as the compounded work of a practical joker (a view which, as we have seen earlier, could not have been entertained by Sir Thomas Browne). When it became clear that the marsupials were real enough, Darwinian evolution became the practical joker, and the following story emerged: Australia and South America were isolated from the great linked land masses of Africa–Eurasia–North America, and so the relatively primitive marsupials carried on in a kind of sheltered workshop, immune from the fierce competition of a free-market ecology prevailing in these more aggressive lands. When a land-bridge was at length established between the Americas, the placental mammals dashed across it, and the backward marsupials of South America were soon displaced by the more efficient placentals. Only in Australia were they able to dodder on.

The biologists who have done research in this field tell a quite different story. It is now thought that reproductive differences had nothing to do with the success of North American over South American mammals in the Tertiary Period. The North Americans would have won out even if they had been marsupials themselves. Their adaptive advantages were in locomotion and feeding characteristics, derived by chance from pre-adaptation – that is, they had already begun to adapt to climatic changes, which were felt earlier in North America than in South America. This does not mean that they were more highly evolved, but only that they were better adapted. As for the Australian marsupials of today, the first point is that they are as 'new' or recently evolved as the fauna of all the other continents. They are not ancient relics at all, but post-Pleistocene, a response to climate change, as every-where else. Only Africa and Asia retained a substantial representation of the Pleistocene megafauna – large animals – that we had and lost; but of course even so the elephant and giraffe and camel are not Pleistocene but Holocene species. They are just a little closer to their ancestors. Our marsupials are less so.

Moreover, the marsupial fauna of Australia is a miracle of adaptation to an environment that is mostly arid and has very high climatic variability. The amaz-ingly efficient production line of marsupials may involve the presence of three sequential stages of offspring associated with their mother at the one time. An almost-weaned young kangaroo (a 'joey') may be still going back occasionally to the pouch to suckle, when a newly arrived diminutive sibling can already be in the pouch, while a very early embryo is waiting ('embryonic diapause', a physiological block to development) in the uterus for its turn to grow.

Other evidence that life in the pouch is not a second-class solution for 'unsuc-cessful' primitive mammals can be found in the degree of sophistication of marsupial milk production. When both older and younger joeys are feeding in the same pouch, the mother is producing two different types of milk: one richer in proteins secreted from the nipple where the younger joey is feeding, and one richer in lipids for the nipple used by the older joey. A four-star catering performance, which makes lactation by 'higher' mammals look like a fast-food line (Giorgi, 1989).

Conclusion

Whether or not there is a world out there independent of our perceptions of it, we cannot escape the variability of those perceptions. The ways in which we perceive, imagine, conceptualise, image, verbalise, relate to, behave towards the natural world are the product of cultural conditioning and individual variation.

One of the rewards of the study of philosophy is to strengthen our defences against abstruse theorising: we should be searching first for clarity. The important question usually is 'what do we mean when we say . . .?', or 'where have these words been before, and what aspects of our past do they trail behind them?' It is important because linguistic structures are conservative, and they pattern our think-ing. The feminists have taught us that we need to struggle to escape sexist language, but that is only one example of cultural bias perpetuated by language. The example of Eurocentrism is another, but only one of many. Our ethical and our aesthetic pronouncements are probably the most suspect from our present point of view.

For the purposes of this discussion, our current intellectual dilemma – one that has many practical consequences – is that we conceptualise 'Nature' in three ways, ways that are not mutually compatible in logic. They can be crudely located along an axis measured off by degrees of internalising and externalising Nature. At one pole, we see ourselves as a part of Nature, a concept at which we arrive through evolutionary theory. Our species is collapsed into the biosphere, rather than set outside it. At the other pole, natural systems are seen as self-regulating and self-maintaining, with our species seen as of very minor significance in the scheme of things; our pretensions to externality, responsibility and some degree of control are irrelevant. The more common forms, however, are those implicit in much ecological writing, which is preoccupied with 'natural systems', 'the balance of Nature', and so on. Human intervention is conceptualised as *disturbance*, almost always seen in a

13

negative light, which clearly externalises us from Nature itself. We operate *on* Nature.

Somewhere in the middle, very uncomfortably sited, is the most common conceptualisation, of our species as a part of Nature, yet at the same time responsible for managing it. The paradoxes of conservation arise from this uneasy compromise – wilderness areas, for instance, are managed to protect them from *disturbance*, human intervention, but how? – by human intervention, of course. And, for whom? Well, for us; for Natural Man rather than Techno Man, but there is still only Us.

References
Blainey, Geoffrey, 1988, *The Great Seesaw: A New View of the Western World, 1750–2000*, Melbourne: Macmillan.
Blake, William, (1793) 1961, 'A Memorable Fancey: Marriage of Heaven and Hell', *Poetry and Prose of William Blake*, London: The Nonesuch Library.
Brewer, D. S. (ed.), 1972, *'The Parlement of Foulys' by Geoffrey Chaucer*, Manchester University Press.
Browne, Sir Thomas, (1642), 1972 edn, *Religio Medici*, Oxford: Clarendon Press.
Foucault, Michel, 1970, *The Order of Things: An Archaeology of the Human Sciences*, London: Tavistock Publications
Giorgi, Piero P., 1989, 'What's Wrong with Marsupials?', in Giovanna Capone (ed.), *Australia and Italy: Contributions to Intellectual Life*, Ravenna: Longo Editore.
Gould, Stephen Jay, 1980, 'What's Wrong With Marsupials?', *New Scientist*, 2 Oct., pp. 27–8.

2

Words and weeds: some notes on language and landscape*

It was Oscar Wilde, I think, who began a talk in the United States with these words: 'We have between us the barrier of a common language'. It has wider application. We stuff the skin of words with the meat of meaning according to our own experience and character. No language is common to two people: the best one can hope for is that there be a fair level of shared meaning. Most misunderstandings arise from the assumption that the same word or phrase means the same thing to different people. There is no solution to this age-old dilemma. The best one can do is to be linguistically aware, but not merely to improve our communication skills. There are bigger things at stake. In this essay, I look at some of the cultural baggage that landscape words and phrases carry around with them. The cultural baggage may imply values and endorse power relations that, consciously, we reject.

These points are true of language in general, and their significance has been given prominence by the women's movement and by underprivileged groups whose status has been confirmed and perpetuated by the common language of the mainstream power group. Words like 'wog', 'nigger', 'poofter', or 'old woman' for a male fusspot ('old women' are always male) are rejected because they encapsulate value judgements and lead to actions which society no longer accepts. The cult of the 'politically correct' has led to excesses that are ridiculed by rednecks but there is no turning back. The language of landscape, however, has not been under such scrutiny, although it is as much in need. Think, for instance, of the different impact on perception and resultant behaviour of the two phrases 'reclaiming foreshore land' and 'filling in part of the river'. We reclaim lost property – it was ours all along. How very convenient for the engineers. For terms in need of a little 'ethnic cleansing', consider the popular and widely used names for *Xanthorrhoea preissii* and *Kingia australis*: 'blackboys' and 'blackgins', respectively. I prefer 'grasstrees'. For an indication of power relations, although ancient, take 'heathen'. Christianity,

*First published as 'Words and Weeds: Some Notes on Language and Landscape', *Landscape Review*, Canterbury, New Zealand, no. 2, 1995, pp. 3–15.

introduced by the Romans to Britain, was a metropolitan religion. The indigenes survived, not in the towns of the valleys, but in the bleak hills and moors, the heathlands, and were thus to be despised. They were the heathen.

The main points I wish to make are as follows:

- Language influences our perception of the environment.
- The words we use reflect our objectives and interests in the environment.
- Language affects our actions in the environment.
- Linguistic awareness is essential to self-awareness; if it is well developed we can modify the way we see the environment and act in and on it.

Anthropocentrism

The language of landscape is inescapably anthropocentric: inescapably, because the very concept of landscape is anthropocentric, a way of positioning ourselves in relation to the external environment. Indeed we are constantly translating that environment into landscape to make it humanly habitable. My point is not, therefore, to urge us to avoid such language, but rather to be more highly conscious of it. To this end, I begin with some obvious examples of explicit anthropocentrism, progressing to more subtle, implicit ones, concluding with a survey of some of the verbal tactics we employ in coping with strange environments. One that is prominent in Australia and New Zealand is the range of naming strategies we use for places and things, a theme that I introduce later (see p. 24).

The point of being highly conscious of the anthropocentric implications of our language is that if we are not, we are servants of the language we use, rather than its master or mistress, and as a consequence, we have little insight into the nature of our perceptions, and limited control over our actions.

'A weed is a plant out of place' – that is a common definition: the *Oxford English Dictionary* puts the same point in slightly different words: 'wild herb springing where it is not wanted'. That restricts it to herbs, but any gardener knows that tree and shrub seedlings can be prime weeds: sycamore seedlings are worse than any dock or thistle. Both definitions are openly, unabashedly anthropocentric. The common weeds do have certain characteristics as plants: they are opportunistic, favour disturbed ground, germinate easily, and have abundant easily dispersed seed of long viability; indeed, from an ecological perspective they are usually the first stage in a succession, the primary colonisers. But this is not what makes them weeds. What makes them weeds is that they spring 'where they are not wanted'.

Yet the wanting can change. Rye and oats began their lives as weeds in cultivated wheat, and in the Americas, tomatoes did likewise in maize. But the soil and climate, the conditions of cultivation, changed as the crops extended their range. In the case of wheat, this happened as it moved north or to higher altitudes. The weeds – rye and oats – grew better than the wheat until in time *they* were recognised as the crop (Tannahill, 1973, p. 30). The tomato came into cultivation in the same way; once a weed, it has become a crop in its own right, but volunteer seedlings in my garden can still be weeds, to be weeded, and it is I who decide. My choice of words is also a choice of actions.

Anthropocentrism is seldom so direct and unambiguous. Much more common is that rhetorical trope known as 'the pathetic fallacy', the projection onto the landscape of human values and emotions. Often this is a conscious projection. If we

CSIRO research on kangaroo grass (*Themeda australis*) near Canberra, ACT. The native kangaroo grass was once common on the Limestone Plains. Its tall stalks, triangular ears and russet brown colours dominated the summer landscape, with spring colour coming from the herbs and daisies that associate with it in this part of Australia. Urban expansion, intensive grazing and the use of superphosphate to promote the growth of legumes have caused it to disappear from much of the ACT; in effect, it became a 'weed', suppressed in favour of the legumes. In the 1980s the Department of Capital Territory proposed that the natural grasslands be retained and regenerated as an important characteristic of the city's landscape and its natural setting. COLIN TOTTERDELL

say that the weather is threatening, we merely mean that it looks like rain; if we describe the sun as malign or pitiless, it is not in the expectation that one day it will have pity, but merely that we find it unpleasantly hot – unless, of course, we are animists, and do really believe that the forces of Nature are directed at our welfare or its opposite. But in the rationalist world, even to talk of a landscape as indifferent – as Thomas Hardy so often does – is to indulge in the pathetic fallacy. To say that a landscape is indifferent to our presence has no meaning unless there is an imagined possibility of its being otherwise. In rationalist terms, the Universe is neither aware nor unaware of our existence; it is simply the Universe.

Because an animistic past underlies the history of the pathetic fallacy, its use is often unconscious rather than conscious, and I find these examples interesting (see also Chapter 13, my essay on May Gibbs, Cuddlepie and other surrogates). It is evident in much description of the Australian outback, which is often described as 'harsh'. It is easy to bring to mind images of David Attenborough in Patagonia, the

Kalahari or the thermal springs at Rotorua, waving his hands at us from the box in our living rooms, enthusing about the capacity of some plant or animal to survive 'in this harsh environment'. The truth of course is that the plant or animal is well adapted to that environment – otherwise it wouldn't be there – but would almost certainly die if you moved it either to your back garden or living room, which *would* be harsh for it.

Another word that shows a similar unconscious anthropocentrism is 'infertile', used of soil in, say, the Stirling Range National Park in Western Australia, which has considerably more species in the one park than there are in the whole of the United Kingdom. The soils are not 'infertile'; indeed even to say they have 'low nutrient status' is loaded, though preferable. What we mean is that the soils are not much good for growing the things we want to grow. They are perfect for growing the things they want to grow – and many gardeners in Victoria who try to grow *Banksia coccinea* or *Dryandra formosa* may wish they had such splendidly productive dirt.

In avoiding the pathetic fallacy, however, I have introduced a close relative, although one from a different discourse: teleological language. By saying that the soils of the Stirling Range National Park are 'perfect for growing the things they want to grow', I may appear to be attributing purpose or design to the soils themselves. Scientists, and especially evolutionary biologists, generally eschew teleological language imputing purpose or design to natural processes. In the present case, the turn of phrase I have used is merely playful, and not likely to be taken seriously, but teleology has also been creeping back into the fringes of scientific discourse, notably in concepts like 'the Selfish Gene', and the Gaia theory of the Earth, which tend to read a set of homeostatic mechanisms as quasi-purposeful devices of self-maintenance on the part of the Earth itself. Moreover, the words you choose to describe these phenomena indicate where you stand – 'mechanisms' on one side, 'self-repair' on the other.

Order and disorder

There is an intriguing group of landscape words that indicate *contrived disorder*, which is, of course, a contradiction in terms, but a deep-seated one: words like 'spilling', 'cascading', 'tumbling', 'drift'. To spill something is to lose a liquid from its container by accident, and there is nothing good about it. No one wants to spill a glass of claret over the best Persian carpet or the hostess' white silk dress. Yet plants 'spilling' over a path are eminently desirable. Why do we use this word, and the others like it? Note that they all have fluid overtones: mountain streams tumble and cascade; a boat on the river may drift aimlessly in the current. But when we aim to plant things in drifts, are we aimless? Not at all. We merely want to appear so. The effects we seek are wholly conscious, highly contrived, but the contrivance should not be apparent. If we have a straight brick path, we *choose* plants to spill out over it, that is, to have an accident. *Ars est celare artem* – the trick is to conceal the artifice.

But why? What is this deep dichotomy in our culture? After all, order and explicit, intentional control are central to the great Italian gardens. Nothing could be more explicit than the regular geometric bed defined by clipped box. There is no question about who is calling the shots. We are. Nature is subject. This set of values is still very strong in current garden literature and performance. A well-tended garden is meritorious; neat edges, a good display of colour, tidy beds, a well-maintained lawn, all bespeaking obvious effort and intervention. The degree of intervention is the measure of the merit, of moral standing and good citizenship, in short, of the kind of person to whom the banks lend money. Yet the effort behind the so-called 'informal garden' is no less. The following is typical of much Australian garden writing in the up-market newspapers: the title is 'The *Right* Way to Make a Bed' (my emphasis).

> Successful flower beds should have a sense of harmony and should provide sufficient internal contrast to make them interesting. There should be enough structure in the bed to place the flowers in context while simultaneously framing them and accentuating their colour.
>   Beds should be planned to provide a continuity of interest throughout the year. (Maddocks, 1994)

Note all the imperatives, the things you *should* do. If you thought you could do what you like in your own garden, you are very wrong. There are rules, and you'd better obey them or you'll be in deep trouble. The emphasis on 'continuity of interest throughout the year', at least as commonly interpreted, is an attempt to deny the seasons of the temperate zone. Only in the wet tropics or in Paradise do the flowers bloom the year round.

The opposing impulse, the love of the unconstrained in nature, has also been a recurrent theme for much of recorded history, a romantic counterpoint to the love of visible order. It has been a near dominant mood in the great romantic periods; at others, a passing grace-note: 'There is a sweet disorder in her dress' says Herrick in the seventeenth century – but just enough disorder to heighten his sense of order. Torn and dirty clothes on a peasant girl mucking out the pig pen would be disorder, but not sweet. There has been a swing in the social status of the two stances in the last two hundred years or so. The pride in visible order in the garden is substantially lower-middle class in Australia; the rambling, spilling, cascading gardens of drifts and 'dear little plants like the thymes' popping up between the crevices of the paving stones – these are substantially upper-middle class, although of course moving down. Hawthorn and Kew and Camberwell in Melbourne are full of (so-called) 'cottage gardens': nearby Richmond has, or used to have, much tidier gardens, without a spill to be seen.

Cultural origins of the order–disorder dichotomy

Why these cultural choices, which are especially well displayed in Australasia, but by no means restricted to it? Remember that both may represent a high level of input – those dear little thymes popping up were all bought in a nursery, the paving was *designed* with planting crevices, they were planted with care, in pockets enriched with Osmacote and peat moss; they need maintenance and, after a few years, replacement. The spontaneity is all illusion. Cultural phenomena like this always need multiple explanations. Clearly sheer fashion is a component, as well as travel to Europe, especially Britain, where the appearance of spontaneity is easier to achieve than it is in much of Australia, at least with the kind of plants that people associate with 'cottage gardens'. But it also seems to me that the value placed on order depends on the degree to which it can be taken for granted. The professional couple in Hawthorn knows that things are going to work, and when not, they can

19

be fixed and paid for. Leaves may get in the gutters: we buy some gutterguard and have it installed, or the gardener who comes once a week can clean the gutters. Leaves in the gutters is the price of trees. Our lives are highly structured. We live by the clock and the appointments book. We want our gardens to look relaxed and carefree, and we will pay for the work that makes them appear so, or buy machines that make it easier. We do not need to *display* the orderliness of our lives. This is a reversal of the baroque garden, with its box hedges and formal *allées*, all on a grand scale – only the rich could afford the army of gardeners to maintain such conspicuous order. But their suburban heirs today are not the lawyers or the doctors, but the bank clerks or the middling civil servants, who got where they are by hard work, as their gardens declare. There is also a colonial component in all this, especially in Australia, where respectability was precious to those emerging from a convict past, and the bush was the enemy, to be tamed (a theme taken further in Chapter 17, The Australian backyard).

Keeping company with the bank clerk, we often find the landscape architect. Michael Hough says of Ottawa, Canada's show piece, that 'much of its designed landscape negates rather than enhances this sense of identity' (1990, p. 114). Elsewhere he puts one of the most disturbing questions landscape architects are likely to face: 'Why are the abandoned but naturally regenerating landscapes one finds everywhere behind the formal civic landscapes of the city so much more interesting than the ones designers are taught to admire and create?' (Hough, 1990, p. 1).

Hough, however, does not succumb to the romantic attitude that wishes to conceal our interventions in the natural scheme of things, the driving force behind the slogan 'Design with Nature' (the title of a book by Ian McHarg; see Chapter 20, The rhetoric and ethics of the environmental protest movement). It is perhaps corrective for a society that has seen nature as the enemy to be told to see it as a friend, but it is neither. If the naturally occurring state of affairs were satisfactory to our purposes, we wouldn't be designing at all, merely picking the fruit in the Garden of Eden. Our designs are always to adapt Nature to our purposes: we build a roof to deflect the rain to where we want it, and not where Nature was directing it. An example that I am fond of is the management of its lagoon by the Venetian Republic. The rivers that ran into it, especially the Brenta, would by now have filled it with sediment. Lagoons are ephemeral. But the Venetians opposed these natural processes by directing the rivers to the north and south of the lagoon to discharge directly into the sea. A mammoth task, and not at all designing with Nature, but rather, *against* it. However, in doing so, they had a fair grasp of *natural processes*. That is the point. In the last 150 years, these natural processes have been forgotten, and despite all our engineering prowess, Venice is endangered, but by erosion rather than sedimentation.

Design with Nature, both the book and slogan, have had a strong influence on the practice of landscape architecture, yet the romantic impulse is often justified by practitioners in scientific or deterministic terms borrowed from ecology, although landscape architects are not noted for their knowledge of this difficult science. What generates this romantic impulse? The moral values we attribute to Nature? Instead of inflating ourselves with scale, in Texan mode ('our mountain is bigger than your mountain'), are we basking in the warmth of nature's moral qualities – or is it an attempt by the profession to have itself taken more seriously by posing as a science?

The history of architecture is littered with foolish slogans and phrases, many of which the 'larkies' (landscape architects) have borrowed from the 'archies' (architects). I always flinch when I hear the absurdly pretentious talk of 'creating spaces' – as if any of us could. At best, we can define them. The jargon of 'design solutions' is also insufferably pretentious, and has no acceptance outside the two professions: before architects and landscape architects can create a design solution, they have to invent a problem where there is none, although there may be an opportunity. However, this essay is not directed towards shallow professional jargon, but towards the ways in which we use words to position ourselves in relation to landscape. The last group of words I shall look at, all still broadly anthropocentric, concern owning, possessing, dispossessing, imprinting, appropriating, scaling, taming, self-justifying, making familiar.

Possessing the landscape

A stunning colonial example is the term 'Crown land': it is mind-boggling to think that nearly all the Gibson Desert in Australia is the property of Queen Elizabeth, along with Hampton Court and Holyrood Palace – at least in name, although the 'Crown' here is really a dummy for the Government of Western Australia. When bits of it are sold off, they are said to be 'alienated', which makes the natural state of things unambiguously clear (the use of this term is an extreme archaism: *alienatio* in Roman law originally meant to transfer property to another, but the term already implied alienation, in the modern sense of being disaffected, in the classical period and was used in this sense by no less a lawyer than Cicero).

Another form of expropriation is to use the landscape to sell things: the Snowy Mountains and the Mackenzie Country in New Zealand to sell tobacco, for examples; the supermarket shelves are loaded with landscape, at least as they are promoted by the television commercials. Even the restaurant menus sell it: you can hardly ever get eggs for breakfast in our better hotels any more, only 'country fresh eggs', while in Perth, practically all beef and most vegetables are 'Harvey fresh', Harvey being a small town down south with distinctly limited production of either. But the promotional use of landscape has a long history. A good deal of the early painting of Sydney Harbour has an arcadian quality, and was consciously or unconsciously promotional, and this is even more marked in Perth, where the view from Kings Park, looking down over the township, allowed the construction of an arcadian image that lasted for 120 years (see Chapter 8, Figures in the landscape), although this was as much self-justificatory as promotional; those who had chosen to come to the isolated little Swan River Colony needed constantly to reassure themselves with this consoling vision, which bore little relation to the reality of their early lives. Paul Carter explores this theme at the national scale in *The Road to Botany Bay* (see Chapter 4, On *The Road to Botany Bay*).

There is a whole suite of verbal tactics for coping with the environment. A common one is to celebrate and exaggerate its scale, hence the rhetoric of the sublime, full of words like 'towering peaks' and 'rushing waterfalls'. It was especially common in the American West, and in Australasian painting and novels of the nineteenth century. A variant is to inflate yourself along with the landscape. The converse is to make yourself secure by scaling down, the rhetorical trope of meiosis, more common in Australia: it lies behind some of our common terms like 'bush', the dictionary meaning of which is 'low shrub', but which we use of 100-metre tall mountain ash forest. A different form of scaling down is to use the language of utility: the Snowy River was often described as a 'wasted resource'. It is now a 'national heritage', although it was a much more powerful river during its days as a 'wasted resource' than it is now as a 'national heritage'.

An example from Henry Kingsley's novel *Geoffrey Hamlyn* uses a combination of the utilitarian rhetoric and that of the Old Testament to legitimise the invasion of the landscapes of East Gippsland:

> 'There are cattle down there, certainly,' I said, 'and a very large number of them; they are not ours, depend upon it: there are men with them, too, or they could not make so much noise . . .'
>
> 'I'll tell you what I think it is, old Jeff; it's some new chums going to cross the watershed and look for new country to the south. If so, let us go down and meet them: they will camp down by the river yonder.'
>
> James was right. All doubt about what the newcomers were was solved before we reached the river, for we could hear the rapid detonation of the stockwhips loud above the lowing of the cattle; so we sat and watched them debouche from the forest into the broad river meadows in the gathering gloom: saw the scene so venerable and ancient, so seldom seen in the Old World – the patriarchs moving into the desert with all their wealth, to find a new pasture ground. A simple primitive action, the first and simplest act of colonisation, yet producing such great results on the history of the world as did the parting of Lot and Abraham in times gone by. (1970, p. 141)

The heifers following the 'lordly bull' and the whole range of patriarchal rhetoric of pastoralism (for example, Stephenson, 1980, *Cattlemen and Huts of the High Plains*) cries out for a feminist perspective to broaden our understanding of these mythic perceptions. This is a complex example because dispossession (of the indigenous people) and possession (by invaders) become fused through invoking the image of Abraham leading his flocks to new pastures, like Moses leading his people to the Promised Land. By using the word 'desert', Kingsley has clearly absorbed the doctrine of *terra nullius*: the land was empty, and the birthright of these intrepid Britons, who took it with God's blessing.

Yet another device is the 'new chum' ploy, a mainstay both of the American Western and the Australian bush yarn: the old hands make themselves feel more secure by casting a recent arrival as incompetent and naive. The contempt for the new chum is good for self-esteem, and functions as a psychological prop. It was often a working-class equivalent of the patriarchal rhetoric of the pastoral culture, which bolsters self-esteem by invoking images of male potency along with those from the Old Testament. Both give off the odour of lonely men in a threatening landscape.

A familiar rhetoric of our own day is 'ecospeak', which, for example, chooses to see certain landscapes as 'wilderness', a pristine landscape unsullied by technology, whereas, insofar as it is true, it is because we have made it so. 'Wilderness' is an artefact, one that became popular when the Americans reached the Pacific Coast, when, in short, there were no 'new' pastures for Abraham and his flocks. William Cronon, writing of New England, presents and dismisses two common perceptions of the impact of Europeans upon the North American continent. The first is embodied in Henry David Thoreau's journal, in which he mourns the loss of a splendid, noble wilderness and its replacement with a 'tamed . . . emasculated country'. The second is the expression of approval for the stages by which savagery has been overcome and civilisation established. Cronon rejects both: 'the choice is not between two landscapes, one with and one without human influence: it is between two human ways of living, two ways of belonging to an ecosystem' (Cronon, 1983). Cronon's example shows that Geoffrey Hamlyn's Promised Land and the 'wilderness' of the ecospeakers are conceptually related; Henry Kingsley's colonisers and the wilderness buffs both invent a wilderness for their own purposes, the first for pastoral invasion, the second for a special and privileged form of recreation.

The self-identification with Abraham is a powerful coping mechanism, in that it goes beyond self-justification – neither the self nor its actions are seen as in need of justification. A similar naive arrogance is shown in the notions of 'exploring' and 'discovery'. An anthropologist writes of Gippsland that: 'Within the tribal territories there existed a network of routes and in some specific places well worn tracks, which the inhabitants used when moving from one place to another' (Gardner, 1992, p. 96) – and the Aborigines followed these when they were acting as guides to 'explorers'. This was true all over Australia. A good example from the western side of the continent is that of Lieutenant Dale pioneering an overland route between Perth and Albany – with, of course, the help of an Aboriginal guide, who conducted him along what is virtually the route of the Albany Highway today, with easy grades and river crossings, fresh water, and abundant food resources, along familiar pathways that had been in common use for hundreds of years. Nevertheless, our history books tell us that it was Lieutenant Dale who mastered the environment by discovering this important route, thus strengthening our right to the land.

Naming

An obvious form of taking possession is to give names to things and places. The field sciences, such as geology, soil science, botany and zoology, all work by giving names that indicate relationships within a set of constructs that are European in their origin, so that this too ties the unknown to the known, reducing unfamiliarity. Consider, for example, the Linnaean binominal taxonomy used in botany and its application to trees like the snow gum of our High Country, or of the southern mahogany from coastal Gippsland (and see discussion on Paul Carter's account of Cook's and Banks' naming practices in Chapter 4). The snow gum is *Eucalyptus niphophila*; the southern mahogany, *Eucalyptus botryoides*. Both are members of the Myrtaceae, the myrtle family, of which the best-known European member is the classical Greek myrtle, sacred to Aphrodite. Thus the application of this taxonomic system immediately ties these two strange and unfamiliar trees into a known world, and not just of science, but to a culture with two millennia of human history. The names themselves are also of Greek construction: *Eucalyptus* means 'well-covered', referring to the operculum (Snugglepot's hat) that covers the flower bud, while *niphophila* means 'snow lover', and *botryoides* means 'bunch of grapes', referring to the way in which the fruits or gum-nuts are clustered together.

That the bestowing of names is a form of possession is beautifully illustrated in Peter Carey's *Oscar and Lucinda*. Carey writes of a piece of celluloid Oscar used as a psychological prop in crossing the seas to Australia in 1865: Oscar's celluloid was perhaps the first synthetic long-chain hydrocarbon in the southern hemisphere.

> This was something my father, being a chemist by training, pondered over, but only once out loud. My mother would not hear him speak of it – When my father spoke of the scientific history of celluloid (which, having a diploma in industrial chemistry, he was entitled to do) she felt that he was contesting her ownership of its original use, its meaning, its history.
>
> And she was right. When my father said 'long-chain hydrocarbon', he was saying: 'I am right. This one's mine.' But my mother would not let him have it. The celluloid was hers. The meaning of it was hers. (Carey, 1988, p. 43)

The point is made even more simply by the use of the word 'granite' to describe the most common rock of the Kosciusko massif and the Monaro. The word derives from the Latin by way of Italian, and means 'granular' – the grains of the minerals of which the rock is composed, mostly feldspar and quartz, can easily be distinguished with the naked eye. The geologists and field naturalists who first saw it around the upper Snowy already knew it well; indeed it is one of the earth's most common rocks. They had probably known it in Scotland, perhaps also in Scandinavia, and if not, at least as a building stone used for half the banks in London. And they knew also what kind of rock it was – it is formed under heat and pressure beneath the earth's surface, unlike the eruptive volcanic rocks. The simple act of naming ties these rounded and polished boulders of our High Country to a whole world of prior experience and, because it ties the unknown to the known, can reasonably be described as a coping mechanism for dealing with the unfamiliar.

To say this is not at all to depreciate it. This is how we make patterns of experience, although it is worth noting that the strongly Eurocentric bias can and sometimes does lead to a misinterpretation of the new, and that the process as a whole was one of expropriation, by which Europeans laid claim to the rest of the world. The new, moreover, was quite often literally taken captive. Aboriginal grave-sites were raided for skulls to be sent back to the British Museum; Bennelong was sent to London, live (but he soon died), while plants were sent back, not only to Kew, but to British nurserymen, to the extent that some newcomers to Sydney Cove in its founding years saw plants growing naturally that they could already identify. They had seen them, not only in illustrations, but alive and in flower in England, propagated by Loudon's nursery from material collected by Banks at Botany Bay. Colonialism has many faces.

Naming places

The naming of places more obviously runs up the flag. Paul Carter goes further: he argues that the 'explorers' did not record objectively the land before them: rather they invented places to conform with European needs and expectations. The naming – which was in fact a renaming – gave substance to their inventions.

> Explorers were not despatched to traverse deserts, but to locate objects of cultural significance: rivers, mountains, meadows, plains of promise. They had a social responsibility to make the most of what they saw, to dignify even hints of the habitable with significant class names. They were expected to arrest the country, to concentrate it into reversible roads which would summarize its content; they were expected to translate its extension into objects of commerce. (Carter, 1987, p. 56)

Many colonial names are about possession and dispossession. Western Australia, for example, is strewn with names like Carnarvon and Albany, Northampton,

Derby, Cambridge Gulf, Perth, Rockingham – all of them singularly unlike their namesakes in Britain. Many of these names displace the names given by earlier migrants. Some of the earlier names survive, and generally seem more appropriate, even though their Aboriginal meanings are lost: names like Meekatharra and Koolyanobbing, Widgiemooltha and Wyalkatchem. There is a further vernacular form of appropriation: those who know Meekatharra, for example, and claim an emotional stake in it or, at the least, familiarity, shorten it to Meeka. In the same way, Fremantle becomes Freo, and Rottnest, Rotto.

The renaming substituted names that were often trivial, recording an imagined likeness to some known place in Britain, or the name of some associate who would mean little to the future inhabitants of the district and nothing to the old, for names that were of profound meaning to the Aborigines who had used them for thousands of years (Watson, 1984, p. 27; Lennon, 1992, p. 151). Thus along the Snowy River, the Tongaroo became the Jacobs River, and the Moyangul, The Pinch.

There is now a preference in the Snowy catchment for the Aboriginal names, although many maps carry both names. Some anthropologists advocate such restitution generally, which may contribute something towards assuaging our guilt complex, but adds little to meaning, since the significance of most of the Aboriginal names has been lost. Their remaining function is well put by the anthropologist Peter Gardner:

> The essence of the Aboriginal names is that they delineate clearly the former Aboriginal occupation of the land and the depth of their association with that land. But for the names, many current residents would deny the prior existence of, and occupation of the country by Aboriginal people. Despite the calamitous and brutal depopulation of the alpine Aborigines the names add a richness, and an immense depth, to an otherwise shallow and chronologically recent culture. (Gardner, 1992, p. 97)

It is not so much the culture that is chronologically recent as its representation in this continent, but Gardner's point is sharpened if we contrast the density of the Aboriginal naming of local features in this, as in most other regions of Australia, with the density of our own naming. Our names are scattered thin on the map, whereas the Aborigines had names for every feature. Gardner estimates that 'probably more than 90% have been irretrievably lost' (1992, p. 97). Thus the net in which we have landed our catch is very coarse, where theirs was fine.

Place names are a measure of the imaginative possession of a region. Some names, used too often, indicate a failure of the imagination: Sandy Point, Long Point, Boggy Creek, Rocky Creek, Sandy Creek, Five Mile Creek. Some are standard all over the English-speaking world: Sugarloaf for a flat-topped hill, The Devils Backbone for a serrated ridge (the Snowy has two of these, one near Campbells Knob and one near Lochend, while there are two Little Rivers and five Sandy Creeks). They are perfunctory names that fail to individuate, which is the function of a name. Others, however, are vividly descriptive, particularly names like Mount Seldom Seen, east of Gelantipy, because it is, and The Washing Machine, a vicious whirlpool rapid. Another rapid aptly named is the A-Frame, made by two huge brick-shaped boulders leaning against each other with the bar of the A formed by the water's surface, funnelled through the narrow opening. Gentle Annie, however, also a vicious rapid, is named by the contrary principle, an understatement that is almost an endearment, the respect for a good adversary combined with laconic wit. All these names show a living human response to a place.

Names also record experience and use – they tell a story. Slaughterhouse Creek and Stockyard Flat are cattlemen's names; the Nine-Mile Pinch is a drover's, redolent of sweat and effort. Dynamite Creek is a miner's name, and so, I suppose, are nearby Dysentery Creek and Bare-Arse Creek. 'Sandy Creek' and 'Five Mile Creek', by contrast, bespeak a weary admission of failure to differentiate. 'One bloody creek after another, and you can't tell 'em apart.' The Aborigines found this response incomprehensible. Thus the intensity and quality of the relations of a culture with the land can be read off the maps, from the density and vitality of the names. There is often poverty in our naming, which betrays the lack of an intimate relation between ourselves and landscape in much of Australia.

Pre-industrial and post-industrial naming

The point can be extended to cover all our relations with the natural environment. In pre-industrial society, country people in England, as elsewhere in Europe, had an intimate knowledge of their environment: they had to. Agricultural workers, for example, had a very large vocabulary enabling them to draw complicated distinctions between different kinds of domestic animals. All pastoral peoples have a large vocabulary for this purpose. Lapland herders are said to have some fifty names for the colour of their reindeer. Hunters had a minutely detailed vocabulary for their prey, distinguishing them by their age and sex and naming their tracks, their cries, their anatomy, their droppings and their behaviour. There was a rich array of collective nouns for each species, a few of which survive today to be aired on quiz nights. Much knowledge of the natural world could thus be taken for granted. It may be that the origins of this knowledge lay, as the anthropologist Claude Lévi-Strauss has suggested, in the universal desire of all people, however 'primitive', to know and classify their biological environment, simply for the sake of knowledge and for the satisfaction of imposing some pattern upon their surroundings. As one observer writes of the cattle-herding Dinka of the Sudan, 'the imaginative satisfactions provided by their herds are scarcely less important than the material benefits'. In early modern England, similarly, the popular nomenclature of plants, birds, beasts and fishes was more elaborate than purely utilitarian considerations required; and much of it had symbolic or emotional value (Thomas, 1983; see also White, 1954).

All this is familiar enough, but there is no Australian equivalent. The Aborigines may have been on equally intimate terms, and so were a few colonial bushmen, but this intimacy has never become a part of popular culture. The so-called 'popular' names of the flora and fauna of East Gippsland, for example, such as apple box and water dragon, fruit pigeon and river peppermint, are little better known than the scientific names, and since the latter are more precise, it is easier to use them, but the price is to distance us from the natural world. To insist on the popular names is to fake an intimacy that is not really there.

Yet intimacy, knowledge, love, the attributing of value are the foundation of all real conservation, and surely also of all good landscape design. They are not always characteristic of technical experts in a remote capital city. The attributing of value, as the English examples show, could be utilitarian, and often was, but it always required close familiarity; and it could also be symbolic, associative, or literary. Probably most Australians have an emotional bonding with the magpie and its carolling ode to dawn. The swan, kangaroo and emu are at least heraldic beings, and symbolic, up to a point. The smell of gum-leaves perhaps has that immediacy of response that characterises the English experience of their natural world, but there are around six hundred species of eucalypt, and they remain largely undifferentiated in the public imagination.

Conclusion

Better naming or changes in verbal behaviour will not solve these problems, although good names can celebrate places, can form the ligaments that tie us to places. The claims I am making, however, are that the way we use words tells us a good deal about the way we relate to landscape – and that a little weeding in the garden of words can help to maintain linguistic health. The language of landscape, like all language, is loaded. The words we use both reveal and influence our perceptions of the environment, reflect our objectives and interests, and affect our actions, including the way we design. Remember Peter Porter's advice: 'Don't touch that word: you don't know where it's been'. You can at least try to find out. The alternative is to languish in semantic prisons of neither your making nor your choosing.

References   **27**
Carey, Peter, 1988, *Oscar and Lucinda*, St Lucia: University of Queensland Press.
Carter, Paul, 1987, *The Road to Botany Bay: An Essay in Spatial History*, London: Faber & Faber.
Cronon, W., 1983, *Changes in the Land: Indians, Colonists and the Ecology of New England*, New York: Hill and Wang.
Gardner, P. D., 1992, 'Aboriginal History in the Victorian Alpine Region', in Babette Scougall (ed.), *Cultural Heritage of the Australian Alps*, Canberra: Australian Alps Liaison Committee.
Hough, Michael, 1990, *Out of Place: Restoring Identity to the Regional Landscape*, New Haven: Yale University Press.
Kingsley, H., (1859) 1970, *The Recollections of Geoffrey Hamlyn*, Hawthorn, Victoria: Lloyd O'Neil Pty Ltd.
Lennon, Jane, 1992, 'European Exploration', in Babette Scougall (ed.), *Cultural Heritage of the Australian Alps*, Canberra: Australian Alps Liaison Committee.
Maddocks, Cheryl, 1994, 'The Right Way to Make a Bed', *Australian*, 'Weekend Review', 2–4 April, p. 14.
Stephenson, Harry, 1980, *Cattlemen and Huts of the High Plains*, Melbourne: Graphic Books.
Tannahill, R., 1973, *Food in History*, Ringwood: Penguin.
Thomas, Keith, 1983, *Man and the Natural World: A History of the Modern Sensibility*, New York: Pantheon.
Watson, D., 1984, *Caledonia Australis*, Sydney: Collins.
White, T. H., 1954, *Bestiary: The Book of Beasts*, London: Jonathan Cape.

3

Journeys through a landscape: Richard Woldendorp's Western Australian photography*

To hear Richard Woldendorp talk about his art as a photographer is to get a subtext for the images. Of course they speak for themselves, but he wants, also, to speak for them, to give their context, their meaning for him, to plot the journey. I think that his own words express well the honesty, clarity and passion of his perception of landscape, but he distrusts his command of English. At his request, these are therefore *my* words, but *our* ideas, the outcome of talking together.

His journey began in a small, flat, low country bordered on one side by the cold grey North Sea, and on the other by a large and powerful neighbour. Holland has often been invaded by both. His people are clean, industrious, very middle class for the most part, and they often describe themselves as phlegmatic, unimaginative, dull. Yet Holland has produced some great painters, and, especially, painters of landscape, and much of the best work is characterised by extreme clarity, both inside (the famous Dutch interiors) and out. Even Rembrandt, who often used chiaroscuro to dramatic effect, still remains sharp, clear, *focused*. Landscapes that are veiled in mist, in 'soft focus', as the photographers say, are not common among the Dutch painters. Another element common to Dutch landscape painting, at first sight surprising, is a sense of space. Holland is very small and very densely populated, but it is flat and, like all the flat lands, it has great skies. In an undulating terrain, the valleys can be claustrophobic, but on the plains, the horizon is not bounded.

Richard Woldendorp is Dutch and still feels Dutch. He is also deeply Western Australian, and at an obvious level, these photographs are both a celebration of and a plea for our sometimes ravaged land. Beyond that, he wants to communicate his vision of this land. That is not so easily put into words, but it is what matters most to him. In his own words, camera and film technology are now so good that almost anyone can take competent photographs, and thus technical skills are declining in importance. Sharp images are not enough. What makes good work stand out from the ruck is the vision. Not just a good eye, although it begins there. A photographer's

*First published as 'Journeys through Landscape', *Westerly*, vol. 36, no. 4, 1991, pp. 55–62.

Low-lying sandy coastline north of Perth, WA; the crescentic dunes behind the mobile dunes indicate a prevailing wind direction at the time of formation. TONY SHEPHERD

vision must be tracked through the choices he makes. Some are more or less technical: choice of camera, lens, focus, film, printing paper. Then there is the choice of subject, vantage point, time of day, time of season, the light, the frame, the composition and, after printing, perhaps cropping – and there is a choice of sequence in the way the images are shown, with the possibilities of counterpoint, contrast, or iterative climax.

What choices does Richard Woldendorp make? The clarity of the images is one of their most striking features, a clarity that comes from sharp focus and from the clarity of the light, but also from clarity of perception. There are no hesitancies or doubts about these images. There is also a clarity of composition, a lack of visual clutter. Pared down, pared to the bone, perhaps – the landscapes are simple, clean, harmonious, almost translucent.

Sometimes they are literally translucent – there are images from Shark Bay in which we see through the water, but cannot see any evidence of the water itself – other than the infinitely seductive colour it generates, that turquoise and aqua-marine of the Indian Ocean along the coast of Western Australia, where the water is pellucid and the sand a fine, white biogenic sand, the comminuted calcium carbonate of a multitude of sea creatures. Much of this coastline is low-lying, and there is no Mt Athos, no towering capes or mountainous headlands, but there are few other places in the world where the colours of the sea itself are so startlingly lovely (see colour illustration following p. 78).

Earlier responses: Dampier, Freycinet, Grey, Arago, Warburton

Many of the photographs are taken from the air, and are thus the record of a special kind of journey. Woldendorp has chosen to interleave his images with words taken from the note books and journals of other, earlier Europeans who encountered these landscapes on journeys of their own. The first of these is Dampier, at Shark Bay in 1699. Then Mme Freycinet just over a century later, commenting – equally unfavourably – on Shark Bay in 1803, which she saw with her husband, Lieutenant Commander Louis de Freycinet. These were sea-voyagers touching land; the last two travellers' tales told here are those of George Grey, who explored the Kalbarri coast by whaler, then the Gascoyne, Shark Bay and Bernier Island, in the late 1830s, but unlike Dampier and the French, he struck well inland; and finally, Colonel Peter Egerton Warburton, whose landscapes were mostly seen from the back of a camel, far from the sea, on his travels through the Northwest and the Great Sandy Desert in 1873.

Only Grey and Warburton had begun to accept the landscape, and they not often. But there are passages of delight in Warburton:

> After travelling six miles, we reached a beautiful clump of large gum-trees, growing in a swamp at the bottom of a small creek, which was hemmed in by a high sand-hill, and then ran through a rocky ridge in which there were fine, clear, deep water-holes 100 feet in circumference. The green foliage of the gum-trees contrasted pleasantly with the red sand-hills on either side, and the barren, rocky ridge in front. Bustard, bronze-wing pigeons, owls, and other birds were seen in the glen, and the whole formed a most gratifying sight after the dreary sand-hill country over which we have travelled. It was a sight which would well repay a few miles' journey in any country. (Warburton, 1886, p. 179)

There is no acceptance in the earlier writers. Driven both by the demands of survival and by centuries of cultural conditioning, they recoiled with horror from the west coast. 'It has been without a single regret that I left that hell on earth, the coast of New Holland', said Rose de Freycinet (in Bassett, 1962, p. 95), and elsewhere: 'I found myself cast upon so horrible a coast without the least resource. My courage forsook me utterly, and I could see nothing but horror about me' (in Dublomb, 1927, p. 8). The point is later driven home by contrast, as they reached the Indonesian archipelago:

> On the 7th, we sighted the island of Rotti, and the following day we were near Simao and Timor. Imagine our satisfaction at seeing the lovely vegetation of these islands. Our eyes were pleasantly rested by this greenery after the sand-dunes and the dry or stunted shrubs of New Holland (in Dublomb, 1927, p. 8)

A good French woman, she enjoyed the Kalbarri oysters, like thousands after her ('decidedly better than those I had eaten in Paris under more comfortable circumstances'), but they alone earned her praise.

A similar physical experience (of the Kalbarri coast) is given spiritual overtones by Jacques Arago, a French artist who accompanied the Freycinet expedition on *Uranie* as draftsman:

There is first an expanse 40 to 60 feet wide beyond the reach of the high tides: then a cliff, partly white as the whitest chalk, partly slashed horizontally with red bands like the brightest bloodstone: and at the summit of these plateaux 15 to 20 fathoms high are seen stunted tree-trunks, sunbaked, shrubs without leaves or verdure, thornbrakes, roots parasitic and murderous, and all this cast upon sand and powdered shells. Not a bird in the air; not a wild beast cry or harmless four-footed thing or murmur of the least water-spring to gladden the earth. Desert everywhere with its cold heart-freezing solitude, and its vast echoless horizon. The soul is oppressed by this sad and silent spectacle of a nerveless lifeless nature, evidently issued but a few centuries from the depths of ocean (in Dublomb, 1927, pp. 5–6)

Woldendorp's camera has recorded such cliffs, 'partly white as the whitest chalk, partly slashed horizontally with red bands like the brightest bloodstone: and at the summit . . . stunted tree-trunks, sunbaked'. But Arago casts a lurid glow over the scene by his word choice; 'slashed', 'bloodstone', 'roots parasitic and murderous', and so on. My own eye sees the colour bands, not as bloodstone, but as the red of iron oxide, so common in this iron-rich province, and the 'stunted' trees are adapted to low rainfall. Woldendorp sees form, texture, the bands of colour making an abstract pattern, and the twisting roots, an arabesque.

Much of Western Australia is desert in the contemporary sense, that of being arid or sub-arid; it is also desert, comparatively speaking, in the eighteenth-century sense of being unpeopled – by the standards of this crowded globe. Arago makes it absolutely so, lifeless, but of course he was wrong. Richard Woldendorp said to me that 'it is as generous as rainforest, on its own terms', and Harry Butler has shown us on television that there is life under every stone. There are miracles of adaptation in this dry country, of which that of the Aboriginal desert people was not the least. Dampier did not think much of them, and said so in an oft-quoted, dismissive text:

31

The inhabitants of this Country are the miserablest People in the world. The Hodmadods [Hottentots] of Monomatapa, though a nasty People, yet for Wealth are Gentlemen to these; who have no Houses and skin Garments, Sheep, Poultry, and Fruits of the Earth, Ostrich Eggs, &c. as the Hodmadods have: And setting aside their Humane Shape, they differ but little from Brutes (Masefield, 1906, p. 453)

But Grey learned otherwise, as did Warburton, when his party:

found a native well with some water, and we soon saw another close by. This discovery caused us immense joy, for we saw the water draining in as fast as we drew it out, and *we thought we had now got the key of the country* and would be able to get water by sinking in any suitable flat. (Warburton, 1886 p. 179, my italics)

The editor adds a note here, as follows:

The native wells, on the discovery of which so often hung the lives of the expedition, and owing to which they were eventually successful in crossing the continent, would hardly come up to an English reader's preconceived notion of a well. They were little holes sunk in the sand with a slight curve, so that the water was often invisible from the surface, and being thus shielded from the burning sun, the evaporation was less, and the liquid cooler. The average depth of the wells was about five feet, though some attained a much greater magnitude. It would be easy to pass within half a dozen yards of these precious reservoirs by daylight and not perceive them, whilst at night their discovery was quite impossible. It is curious to speculate on the instinct that enables the degraded inhabitants of this wilderness to find the few spots where the precious element is attainable. The savage has the advantage of the European in this respect. Out of forty-nine or fifty attempts made by Colonel Warburton's party to find water by sinking, only one was successful, although in the selection of likely spots they brought all their experience and desert-craft to bear. How often, when travelling in the dark, and perishing from thirst, they may have unconsciously passed wells, a knowledge of which would have been as new life and strength to both man and beast, it is impossible to say. (Warburton, 1886, pp. 179–80)

This editor (from the 1880s) had failed to learn what Warburton himself did learn in time: there was indeed *a key to the country*, but it did not lie with the wells; it lay in the skills of the people who knew where and how to make them. The explorers, the successful ones, learned more and more to rely on those skills.

We have since acquired other skills, those of our current technology, which keep the tyranny of distance a little at bay, and we are less constantly preoccupied with sheer survival. Woldendorp's journeys by plane or in four-wheel drive vehicles, with abundant provisions, good communications and a safety-net of maps, rescue services and so on, are very different from the earlier ones, and it is this comparative security that makes possible, or easier, a degree of aesthetic detachment. The security is comparative only: he rarely feels entirely safe, that nagging question 'what if something goes wrong' always there in the mind's attic. Yet Woldendorp does not see this as a harsh country, or as a 'hostile environment', as it has so often been described. It is simply there, with its own beauty. He has accepted it, and the image-making has been *his* key to acceptance, it of him and he of it.

Woldendorp's eye

Woldendorp sees with an innocent eye, but not an untutored one. It has been said that he photographs like a painter. He *was* a painter. The interplay between the art of the photographer and that of the painter is brilliantly evoked in John Scott's *Landscapes of Western Australia* (1986), which works by pairing a Woldendorp image with a painting of the same or a similar subject. The interplay is complex. Sometimes a Woldendorp photograph has been the inspiration for a painting – for example, *Xanthorrhoea* (1982) and *Flood Creek* (1980), both works by Robert Juniper (a close friend of Woldendorp). *Xanthorrhoea* was inspired by a fish-eye lens photograph by Woldendorp of grasstrees, and *Flood Creek* by an aerial photograph of a tidal river across mangrove flats in the Kimberley. But the influence runs both ways. In another pair, *Pilbara Landscape* by Woldendorp and *The Pilbara* (1979) by Fred Williams, the text may suggest to the naive that the photograph is representational, whereas the painting interprets and transfigures. 'I want to isolate those marks, turn them into handwriting. They become an alphabet, like hiero-glyphics', said Fred Williams. Woldendorp has chosen a clear but low sun and an oblique aerial perspective to highlight the calligraphic quality of the ghost gums in exactly this way – a vision in this case almost certainly shaped by familiarity with Fred Williams' paintings of the Pilbara, and yet also a two-way traffic, both recreat-ing the Pilbara until 'life imitates art'. Because this two-way traffic is so imagina-tively displayed in the book, it is doubly surprising to read in Scott's introduction:

> Just as significant perhaps is the fact that in the desert the artist has little to fear from his modern rival, the camera, whose technology seemed to many at first to be better equipped to show things 'as they are'. But that same technology cannot reveal the limitless expanse of harsh terrain, the rugged dryness, the 'bare bones' and – what we look for in art – man's response to them (in Emile Zola's apt phrase, 'a corner of nature seen through a temperament'). In other words, we look at these artists' works in order to see nature *as experienced by man* and not as observed through the lens of a machine. (Scott 1986, pp. 7, 8, my italics)

But a paint-brush is a tool, albeit a simple one, and neither paint-brush nor camera is mechanical in the sense of being autonomous. Woldendorp makes choices, he selects, and through a combination of technical and imaginative skill he is able to select to the point of abstracting a formal essence of the landscape that is his subject. Zola's phrase 'a corner of nature seen through a temperament' describes this process most aptly.

Woldendorp was not only a painter himself: his seeing eye has been shaped by a deep pictorial culture. There is no mimicking here, but we can see that he has studied and learned from David Hockney's sharp definition and simple, tender colouring; from the calligraphic shorthand of Paul Klee and Fred Williams; and, stretching back into the past, from that long Dutch tradition – Vermeer, for example. The content is different, but both Vermeer and Woldendorp face the world with the same unwavering gaze, the same clarity, the same absorption in pattern and texture. It stretches the mind only a little to think of these images of the Northwest or Shark Bay as 'Dutch exteriors'.

Many of the best 'Dutch interiors' have an abstract quality. At first glance, they are meticulously representational, yet the content is not the point, or not the only point. Subject is collapsed into object as the interest is held by the pattern of tiles on the floor, the quality of light as it is reflected off copper pots and pans, the texture of fabrics. Woldendorp's images are abstract in this sense, most obviously in his aerial photography. Often he seems to have built himself a stable platform in the sky to give him the angle he needs, waiting for the light he needs, often a low light that allows him to abstract pure form from the landscape below. In prosaic fact, he taught himself to fly so that he could make such hair's breadth choices.

The nature of his journey and the landscape images themselves are an interaction between what is there and what the journeyman brings with him. This is a very personal record, but it is also a composite one, a search both by Woldendorp and by European man to comprehend and relate to these clear landscapes where no sophistry survives. It is a journey from Holland through the Gascoyne and the Pilbara, bringing a part of Holland all the way; and collectively it is a different journey, a European voyage through time, from New Holland to a Western Australia still evolving through the exercise of the creative imagination.

The process has no ending. We need constantly to be helped to see in new ways because we fall so easily into stock responses. This is a recurring problem with the words that are so often used to describe our landscapes. Out come the clichés: it is always 'vast', 'harsh', 'hostile', 'unforgiving', all thoughtlessly Eurocentric words.

Vast? A square mile in Western Australia is exactly the same size as a square mile anywhere else. No one talks about the vast landscapes of Europe, although

33

Europe with Russia is about the same size as Australia. Western Australia is sparsely populated and it is a large *political unit*, but neither of these facts adds one cubit to the extent of its *landscapes*. Undifferentiated, perhaps? No landmarks? Well, yes: all Chinese are said to look alike, too. Yet they still seem able to pick out their friends and family from the undifferentiated mass, just as the Aborigines knew their territory intimately, and could hardly conceive how one part of the Great Sandy Desert could be mistaken for another. Landmarking is a perceptual skill.

Harsh, hostile, unforgiving? This, of course, is the pathetic fallacy, projecting human sentiments on to the landscape itself. The alternative is to think of the landscape as indifferent to our purposes, or better, conducive to some and not to others. The onus is then on ourselves to determine which. The thoughtless Eurocentrism of these words irritates me. The few unlucky Aborigines who were taken from this – to them, benign – environment, to England, found the miserable cold and ruthless microbiota, the bacilli and viruses of European diseases, intolerably harsh, and died. But then the malarial swamps of the Maremma, or the plains below Vesuvius, or the disastrous floods of the Po Valley, or Florence through the Black Death – all these give the lie to the popular image of Italy, for example, as a smiling land wrought by the hand of God for the uses of civilised man, in contrast with the stark and barbarous landscapes of the Pilbara. 'Harsh' is where you find it. The pathetic fallacy may have some value. Its negative mode ('unforgiving', etc.) may remind would-be travellers through the Northwest to equip themselves thoughtfully, and the positive mode ('as generous as rainforest') might help us to accept the landscape to the point of caring for it adequately. But in general, stock responses blunt sensibilities rather than sharpen them, and that is why we should be grateful for new ways of seeing, which these images give us.

References

Bassett, Marnie, 1962, *Realms and Islands: The World Voyage of Rose de Freycinet in the Corvette Uranie 1818–1820*, Melbourne: Oxford University Press.

Dublomb, Charles (ed.), 1927, *Extracts from the Journal of Rose Saucles de Freycinet*, Paris: Sociétés d'éditions Géographiques, Maritimes et Coloniales.

Masefield, John (ed.), 1906, *William Dampier's Voyages: Consisting of a New Voyage Around the World, etc.*, London: E. Grant Richards.

Scott, John, 1986, *Landscapes of Western Australia*, Claremont, Western Australia: Aeolian Press.

Warburton, Peter Egerton, 1886, *Journey Across the Western Interior of Australia, with an Introduction and Additions by Charles H. Eden*; ed. H. W. Bates, 1982, facsimile edition, Victoria Park, WA: Hesperian Press.

Woldendorp, Richard, 1992, *Journey Through a Landscape: Richard Woldendorp's Australia*, West Perth: Sandpiper Press.

On **The Road to Botany Bay**: a review of a book with that name by Paul Carter*

Reviewing the reviews

Paul Carter's book has created a mild stir. Critical response in the reviews has been highly diverse, or as Elizabeth Swanson puts it in the *Australian* (1987):

> Pity the poor book review reader. What should she think when reading these two reviews of Paul Carter's *The Road to Botany Bay* (Faber & Faber, $29.95): 'Australian history will never be the same after Paul Carter's brilliant, beautifully written book' (Barry Hill). 'Whenever he eschews theorist jargon, Carter writes elegantly, but his 370 pages of exposition are the hardest jungle of words I have ever had to fight my way through' (Peter Porter).

The range of opinion says something about the current intellectual climate in Australia, so I shall review the reviews as well as the book.

Nearly everyone found it hard going, except two novelists, Barry Hill and David Malouf, who is quoted on the dustjacket (source not given) as finding that 'the writing has a lyrical passion in argument that I found irresistible. I couldn't put it down'. Almost everyone else could put it down fairly easily. 'This book is not an easy read', begins Manning Clark (1987), in the most curmudgeonly of the reviews. Christina Thompson's verdict:

> It is by turns interesting and infuriating. At times, you think Carter is doggedly pursuing what he has already demonstrated to more than your satisfaction. At others, you wonder what the hell he's talking about. Sometimes the light clicks on, sometimes it doesn't; sometimes that's your fault, sometimes it's his.

However, she also gives a clue as to why this is so:

> As a book *The Road to Botany Bay* is characterised by its preoccupation with language as a subject rather than a medium. It is also particularly susceptible to the aesthetic delights of the dialectic. Stylistically, it hovers somewhere between a kind of elegant belle lettrism and the more technical and less specific discursive practices of contemporary theoretical writing. (Thompson, 1987)

*First published as 'On "The Road to Botany Bay"', *Westerly*, vol. 33, no. 4, 1988, pp. 15–26.

Even a sympathetic and perceptive reviewer, David Dolan, comes down hard on the style of the book:

> In parts, *The Road to Botany Bay* does seem like a few pungent footnotes enormously bloated into a whole book. Nowhere in his 400-odd pages does Carter tell us clearly and succinctly what he is attempting to do or why; but then he rarely puts anything succinctly. With its half-realised promise of new insights, this pioneering essay in 'spatial history' looks suspiciously like a dead-end in itself: a good idea that fizzled. (Dolan, 1988)

Two well-known expatriate Australian men of letters, Peter Porter and Clive James, both give it stick. Porter says that the book:

> might be described as a fantasy on one note, since 'space' and 'spatial' occur in almost every paragraph. Thirty pages in, I made a note for my later perusal, 'What is spatial history?' Three hundred pages later, I still didn't know.

And a little later, '*The Road to Botany Bay* shows what happens when those who luxuriate in words dispense with the nimbus of common experience which words carry with them' (Porter, 1988). Clive James is more savage: he concludes a review (with the title 'Bullshit and Beyond') aphoristically: 'More briefly, bullshit is empty depth. Mr Carter feels obliged to deploy his chic vocabulary not because his big idea is new but because it is a truism' (James, 1988).

Reviewing the historians

So much for the writing. There is less agreement about the intellectual claims of the book. It bears the subtitle, 'An Essay in Spatial History', and this is where the trouble starts. Since time cannot be conceived without space, nor space without time, this subtitle does not make much sense, and the author does not help us, for we are never told directly what the phrase means to him. Nevertheless he sets it up as alternative, and superior, to what he calls 'traditional' or 'chronological' or 'imperial' or 'empirical' history – 'almost as though these terms were synonymous. Perhaps in his mind they are', says Russel Ward (1987) tartly. Carter's claims to be founding an entirely new kind of history are rejected out of hand by Russel Ward and Manning Clark – quite properly, I believe.

Yet Carter undoubtedly has something useful to say about the way our historians present history. In writing with hindsight, they falsify the experience of their subjects. Stuart Macintyre, one of the few historians who has taken the trouble to find out what Carter really achieves in the book, put the case against his colleagues' work thus:

> all of these efforts to write the history of exploration and settlement assume what they seek to explain. For all of them the country was there waiting to be found [but] the assumption that historical space lay dormant waiting to be found robs it of its historical content.
>
> His [Carter's] own spatial history proceeds from the insight that the place was not there in advance. Rather, he insists, it was brought into being by the very act of exploring, travelling, settling, apprehending. (Macintyre, 1988)

Another way of putting it is to say that it is a book about the way in which space becomes humanised, brought within the realm of our culture, translated into place. Surely Carter is right in saying that school history falsifies this process. We all remember the maps in our textbooks showing the routes taken by Sturt, Eyre, Burke and Wills. Their journeys are plotted on our own map of Australia, one that has now been two hundred years in the making. But for them it was never like that. What lay ahead each day was quite blank, wholly unknown. They unrolled the map with the day's journey. This is true not only of the explorers. Governor Phillip could not know the consequences of moving from Botany Bay to Sydney Cove. Historians can, looking back, or at least they can try; they may attempt to link present consequences to early causes, in a long chain of events, selected and linked – interpreted – in hindsight. Thus, one might say, historians explain the present by falsifying the past, in that *their* past (that is, for example, their 1788) can never be that of the actors. But then surely we know before we begin to read a history book that it is, to use Manning Clark's word, a *story*: one man's interpretation, a sequence of linked events; and, of course, retrospective. That *is* history (that is, historiography).

The Road to Botany Bay, on the other hand, is not history, not a new way of writing history, and although it is in part an intellectual critique of history, there was really no need for Manning Clark to get upset. For Manning weighs in with heavy irony:

> How did Australian historians see the beginning of white settlement in Australia? He [that is, Carter] knows the answer: for the historians, Australia was 'a stage where history always occurred'. History was 'the playwright co-ordinating facts into a coloured sequence'. The historian is merely a 'copyist' and a selector, little more than an editor of the playwright's work. Exit previous historians. Enter Paul Carter to tell us what history really is. (Clark, 1987)

Mind you, Clark was provoked. Clive James offers the comment 'that to call Manning Clark an imperialist historian is like saying that Bertolt Brecht had a crush on the Duchess of Windsor'.

When criticising Clark's *History* in the introduction, Carter tells us that: 'It is not the historian who stages events, weaving them together to form a plot, but History itself'. But a little later, 'Clark's description does not simply reproduce the events: it narrates them, clarifies them and orders them'. Then on the next page: 'The historian does not order the facts, he conforms to them'. Academics are used to marking down such inconsistencies in their undergraduate essays.

The book is full of self-contradictions, and it is mischievously provocative, but Manning Clark and Russel Ward were never seriously under threat. As David Dolan points out: 'Notwithstanding the occasional sniping shots at imperial and empirical history, *The Road to Botany Bay* is not a rival to the histories written by Crawford, Ward, Clark, Crowley or Serle' (Dolan, 1988).

If the book is not a new way of writing history – as the subtitle claims – what then is it? I will give my own answer first, before those offered by some other reviewers. I believe that it belongs to a quite familiar category, applied in a somewhat novel way. It is what in English departments around the land is known as Lit. Crit. – textual exegesis – but applied to primary historical sources such as the explorers' journals, rather than to more familiar literary works such as novels, poems or plays. Like almost all contemporary literary criticism, it is influenced by structuralism and deconstruction, the theoretical positions and analytical approaches exemplified in the work of people like Jacques Derrida, approaches begun by anthropologists, elaborated in Paris, and now international in application in literary circles. 'Adorno, Ricoeur, Deleuze, Benjamin, Derrida: meet Manning Clark', says Christina Thompson (1987). 'Carter's own intellectual culture draws heavily upon French structuralism and deconstruction, which could drive our less intellectual and more provincial historians off their maps', says Barry Hill gleefully (1987), but I suspect that he overrated Paul Carter's knowledge of French social theory, and failed to appreciate that they nearly all talk like this in the English Departments. David Dolan offers a more sober explanation of the differences in critical responses:

> If the historians reviewing *The Road to Botany Bay* have missed the point, it is not hard to see why the novelists have loved it. Suggestive, and subtly nuanced, it concentrates on individual human sensitivities, perceptions, and mental processes. It is not hard going at all unless the reader is determined to tie down every single statement into a progression of logical reasoning from a premise to a conclusion. It is not that sort of discourse. Better to press on, taking it as it comes, letting it add up in your mind. (Dolan, 1988)

What are the rewards for pressing on (and on and on – for it *is* hard going, or I found it so)? First, there are many insights about language itself. English is a European language, and it evolved in a northern environment. The available words did not fit this very different world. In time, streams become creeks, fields become paddocks, giving what were dialect words in English a new currency with new associations, replacing the overly gentle English words. In the Shire of Wingecarribee, near Bowral in New South Wales, one may still find the Medway Rivulet, which seems a quaint survival in a quasi-English countryside. But Sandy Creek, Boggy Creek and Dead Horse Creek are more familiar. These are my examples, not Carter's, who has others. This area of intellectual exploration is not new. There are two good books which explored, twenty-one and eighteen years before Carter, the way in which Australian writers have had to forge a new expressive language: Brian Elliott's *The Landscape of Australian Poetry*, and Coral Lansbury's *Arcady in Australia*. Both prefigure Carter's concern for the intersection between words and place.

Paul Carter is, however, certainly original in the thoroughness with which he scrutinises the naming practices of the explorers, and this is one of the two most illuminating aspects of the book. Secondly, he is also sometimes brilliantly successful in recreating their experience: the experience of all those facing a new and unknown world – unknown physically, mentally and psychologically – with adequate ways of knowing yet to be forged. This is not uniquely original, either. Greg Dening (1980), for example, has shown great insight into the nature of the contact experience between Pacific Islanders and Europeans in *Islands and Beaches*, a book which combines the experiential sensitivities of the anthropologist with the narrative skills of the historian by breaking the story with meditations. So Carter is not creating a new discipline, as a couple of his reviewers suggest, any more than he will replace conventional history with 'spatial history'. As Barry Hill (1987) points out, *The Naming of Australia* would have been a better subtitle, and it would have saved a good deal of fuss.

Carter at work: on Major Mitchell

It is now time to take a few extended examples of Carter at work. Let us first take Major Mitchell, who is the subject of his chapter 4, 'Triangles of Life'. Mitchell was a man with a method.

> As he had explained in his little book, *Outlines of Surveying*, 'The most essential operation, in taking a plan, consists in laying down points representing the true relationship of the most prominent objects on the face of the earth.' In this definition of 'taking a plan', 'prominent objects' were not simply what is mapped: they were also the means of mapping. They not only provided the surveyor with worthwhile objects to plot: they also supplied him with indispensable points of view. The explorer might proceed like a ship at sea, but the surveyor had to tack between definite points of reference. Triangulation was essential to the surveyor's progress. (Carter, 1987, p. 102)

He was a good surveyor, as the testimony of his second-in-command, Stapylton, who did not like him, makes clear: 'Sur. Gl. makes capital way across the country, keeps his line wonderfully well and shows a complete knowledge of his subject.'

But Mitchell seems to have made some serious mistakes, of which the best known is his discovery of the Victoria River, 'a river leading to India, "the nacimiento de la especeria" or region where spices grew: the grand goal, in short, of explorers by sea and land from Columbus downwards' (Mitchell, quoted in Carter, 1987, p. 106). What this means is that he thought he had discovered a big river running to the Gulf of Carpentaria, if not quite to India, and the phrase, which he had picked up during the Peninsular War, invokes the shades of Vasco da Gama, Henry the Navigator, and the poet Camões, whose epic, *Os Lusiadas*, he had translated into English. The river so fancifully described was in fact the Barcoo, one of the headwaters of Coopers Creek, which grinds to a halt in the arid interior. Most of us would see Mitchell's prose as an attempt to inflate the importance of his discoveries, but Carter wants to suggest that there is more to it. I shall quote him at length:

> Mitchell's geographical fantasies were not embarrassing weaknesses, but were, in fact, essential to his notion of travelling. They were ways of rendering the country habitable, the grander project which distinguished the surveyor from the explorer and which justified Mitchell in taking his historical role so seriously . . .
>
> One way of defining the difference between the explorer and the surveyor is to contrast the explorer's desire to constitute space as a track with the surveyor's interest in regionalising it. If, for the explorer, mountains and rivers were both means of getting on, then, for the surveyor, they functioned primarily as natural boundaries. They were geographical givens which helped the surveyor bound a useful space, a space that was conceptually and trigonometrically consistent. As Mitchell wrote in his *Outlines*: 'A ridge or chain of heights affords also the most favourable line for the boundary of a plan, or for joining two plans together. Mountains divide the sources of rivers, govern the direction of roads, and bound the visible horizon.'
>
> Where the explorer saw ranges as roads, the surveyor thought of them as bases. Where the explorer's space was two-dimensional, backwards and forwards, the space of the surveyor was triangular, extending in depth to either side. (p. 108)

Mitchell's account of Australia Felix (south-western Victoria) was also open to attack as gross exaggeration. George Augustus Robinson was particularly scathing about the country around Avoca:

> The Major's Line was through the sweep of this plain and he describes it in glowing terms, his usual practice. He came to a level plain resembling a park, hence he called it Major Mitchell's park. The banks of the river Loddon, which was on the E. side, are abrupt but covered with grass. And the river, he said, was north among some hills,

39

probably to water a country of a fine and interesting character. Now this is all fudge. Better the Major had not published such nonsense as it has occasioned an expedition of time and money to numerous emancipists who have gone in search of this country of interesting character. The Major's Eden is another specimen of his puff: excellence not yet located. Eden though it be, the same fate extends to the greater part of his Australia Felix. (Carter, pp. 110–11)

But this fails to understand the function of Mitchell's rhetoric: 'These beautiful recesses of unpeopled earth could no longer remain unknown', says Mitchell. 'He framed a picturesque discourse designed to open up the country.' 'In short', says Carter in concluding,

> the character of exploration as a rhetorically consistent gesture, a strategic previsioning of occupation, is not to be construed as a personal idiosyncrasy of Mitchell's. He simply maintained a culturally coherent view of his spatial and hence historical responsibilities. (p. 134)

It is useful to be reminded that Mitchell was not alone. Captain James Stirling, and Charles Fraser, the Government Botanist of New South Wales, did the same for Swan River, with the same aim of seeing it settled, much as estate agents today describe houses with the aim of seeing them sold. On the whole I prefer Robinson's good eighteenth-century word 'puffing' to Carter's 'rhetorically consistent gesture', but Carter's account puts Mitchell into cultural context, and it also reveals much by the way, especially the importance of hills, ranges, lakes and rivers, even insignificant hills and dry lakes, if one is to mark the land in a country with few landmarks. This makes sense of the names on our maps of the inland.

Naming practices

Some of Carter's major points will be the subject of debate in the journals. His account of naming practices is nearly always interesting and stimulating, but I question some of it, particularly the contrast he makes in the early chapters between Cook's 'open' naming and Banks' 'closed' naming. Stephen Murray-Smith sums up Carter's position beautifully in a particularly perceptive review. Cook named 'The Three Brothers', for example, because they

> 'bore some resemblance to each other'. The names were not arbitrary. They all had a point. But Cook was not trying to impose a world view on the objects under his scrutiny. His names are not 'definitive statements of arrival' so much as points of departure. They are 'metaphors of the journey'. Experience, Carter explains, is *spatial*. Knowledge is many sided, its linkages infinite. Banks, as a botanist, was interested above all in naming things to dispose of them as he poked them into his vasculum. To do this is to abort understanding and to deny life. Knowledge thus loses all power to signify beyond itself, to suggest lines of development or the subtler influences of climate, ground and aspect. It is the difference between the aridity of *discovery* and the fruitful interaction of exploration. Banks, when he seeks to name things, does so to possess them by pretended objectivity. Cook's subjectivity, his ironic use of language, and in due course Flinders's too, is by contrast an 'authentic mode of knowing', the creation of metaphors which symbolise rather than seek to possess. 'Cook's place names were tools of travelling rather than fruits of travel'. (Murray-Smith, 1988)

This is an arresting contrast and an intellectually challenging one because Carter makes of it a remote, colonial playing out of the great eighteenth-century philosophical debate between Locke and Hume. This, indeed, is the excitement of *The Road to Botany Bay*. Cook and Banks, by those sandy shores at the world's edge, become centre stage in Carter's hands:

> How, for instance, does exploration differ from that other great eighteenth-century naming discipline? In answering this question, the conjunction of Cook and Banks on the *Endeavour* voyage is a particularly fortuitous one. Historically speaking, the distinction between botany's concern to reduce the variety of the world to a uniform and universally valid taxonomy and exploration's pursuit of a mode of knowing that was dynamic, concerned with the world as it appeared, went back, on the one hand, to the Enlightenment project of universal knowledge and, on the other, to the trenchant criticism of its empirical assumptions mounted by David Hume. But nowhere is this methodological distinction brought out more clearly than in the contrast between the specific practices of Banks and Cook. (Carter, p. 18)

And so the theme grows: for where Banks was preoccupied with the typical, Cook was concerned with the singular; where Banks tended to generalise, Cook tended to specify. 'And this, indeed, was the difference between botany and geography as they were practised in the eighteenth century' (p. 18). This argument is then developed more fully a few pages later:

> A profounder distinction between botany and exploration now emerges. For the difference between the two was not simply a matter of methodology: it embodied, more fundamentally, a disagreement about the nature of language and its relationship to the world. For Banks, names enjoyed a simple, Linnaean relationship with the object they denoted. They gave the illusion of knowing under the guise of naming. Cook's names obey a different, more oblique, logic: the logic of metaphor. (p. 29)

An extended quotation from Carter at his most illuminating rounds out this theme:

> Figures of speech, place names among them, correspond symbolically to the scope of exploration itself: they are a means of making sinuous paths comprehensible, a means of recording the journey as it impresses itself on the consciousness. There may be nothing objective about this, but then, as the philosopher Paul Ricoeur has observed, 'There is no non-metaphorical standpoint from which we could look upon metaphor, and all the other figures for that matter, as if they were a game played before one's eyes.' Similarly, there is no non-directional, unimplicated point of view from which the traveller can describe the facts of the journey.
>
> Cook's metaphorical mode of naming is not a peculiar whim of the namer: it represents an authentic mode of knowing, a travelling epistemology that recognizes that the translation of experience into texts is necessarily a process of symbolizing, a process of bringing invisible things into focus in the horizontal lines of the written page. So, where the metaphorical nature of Banks's discourse is suppressed, Cook feels under no such constraint. Names like 'Pigeon House Mountain' or 'Mount Dromedary' spectacularly depend on the namer's point of view. They assert no literal likeness but are offspring of the paradoxical miniaturization of the magnified image in the telescope, framed and isolated, such features are brought close, made homely, domestic. They are grand enough to hang on a wall; small enough to fit into a pocket. But Cook's seaworthy metaphors do not in any way diminish the otherness they make so readily accessible. For implicitly in his metaphors is the figure of irony, a mode of description

that passionately distances the observer from what he sees. If Banks's generalizations tend to belittle the coast, then the particularity of Cook's inventions suggests nothing so much as humility, a willingness to be dwarfed as well as to command. (pp. 30–1)

Kinds of naming

This is good stuff, the kind of writing that led David Malouf to talk about 'a lyrical passion in argument'. Bold, imaginative, exciting – yes – but will it quite do as argument? If Cook represents an 'authentic mode of knowing', Banks by implication does not. Moreover, they are dramatised 'stand-ins', Cook for the experiential, the existential, against Banks, the generalising way of the classificatory sciences, which struggle to discriminate likeness and difference through genus and species, to establish relationships, to build upon objective knowledge of the world. The Artist is set against the Scientist, although the Artist is a product of the Romantic Revolution and not of the eighteenth century. Dr Johnson was quite clear that it was *not* the business of the poet to number the streaks of the tulip.

The critique of the pretensions to objectivity of eighteenth-century science, and indeed of all science, is proper and needful, but the case presented here is surely incomplete. Both modes of knowing have their achievements. Carter's contrast is incomplete in two ways. First, even though I accept his point that Banks's taxonomy seems stultifying, it was in the end going somewhere, because it was part of a collective and continuing scientific endeavour, whereas Cook's geographical names were not.

Cook was naming places. Banks, in naming plants, was tied into a different linguistic system. Plants were named according to a hierarchy, the Linnaean system of classification, and the names indicated presumed relationships. It was not possible to name without suggesting a slot in an ordered scheme, but the system was not closed. It evolved, and was steadily self-correcting in the way of science. Evolutionary theory came to underwrite and modify the Linnaean scheme. Plate tectonics later explained some of the vagaries of distribution, such as the Proteaceae of Africa and Australia. 'Provisional' assignment to genera and families was often reassessed. The systematic Eurocentric biases inherent in a taxonomic system that evolved in temperate Europe were at length recognised. It has even become possible for a reputable Australian botanist (Dr Len Webb) to argue that the origin of the angiosperms may be reflected in North Queensland genera such as *Idiosperma*.

The naming of places is, by comparison, unfettered – one name would do as well as another, or near enough. Nothing much hangs on it. Why? Because the critical act is not the naming at all, but the recording of the geographical coordinates, for which the name is a vulgar mnemonic. There is no flexibility in giving the coordinates, however, and they are a kind of possession, and an imperial one in Cook's case (as he well knew them to be!). The parallels of latitude run from the equator to the poles, and they are a neutral division of space, but the meridians of longitude spread out from Greenwich, the nerve centre of the British Admiralty. Every place named by Cook was also plotted by him on this grid, knitted into the fabric that was woven by the world's then greatest naval power and thus tied to the Britain which he served.

Even so, Carter's book has set me to think about other acts of naming, which is never an innocent act. Most of Melbourne east of the Yarra sits on Ordovician

42

mudstones. To say so is to be scientifically objective – but it is also to invoke much cultural history, and tie the land by words to the Welsh mountains, to Sir Roderick Murchison, Adam Sedgwick, and those remote Welsh tribes, the Ordovices, the Silures, and to Cambria itself, which gave its name to the first of these Palaeozoic sequences. The very sequences themselves have proved both a major tool and a major hindrance in understanding earth history, but once again science has proved self-correcting, and terms that look closed have proved not to be so.

Carter's central thesis

Books that make us think are welcome, and I agree with those critics who have found *The Road to Botany Bay* original, imaginative and stimulating, if also at times infuriating. It is also a substantial work of scholarship in that Carter is both widely and deeply read, with a very thorough knowledge of his primary texts. Stuart Macintyre puts it thus: 'This is not one of those modish attempts to throw a theoretical grid over commonplaces. The material is hard won by the author himself. His analytical grip on the material is sure' (Macintyre, 1988).

That is high praise from a historian. Yet even the favourable reviewers seem to find it difficult to come to grips with Carter's central thesis – perhaps because they take it for granted, while the hostile ones reject it out of hand. There is also a difficulty that Carter imposes wilfully, by adorning his thesis with red herrings. There is, nevertheless, a central thesis, and one that is worth taking seriously. It is perhaps best approached through a passage from Adorno and Horkheimer (1973) quoted by Carter:

> Bourgeois society is ruled by equivalence. It makes the dissimilar comparable by reducing it to abstract quantities. To the Enlightenment, that which does not reduce to numbers, and ultimately to the one, becomes illusion; modern positivism writes it off as literature. (p. 211)

But, Carter implies, literature is not 'mere' at all: *it is an authentic mode of knowing, and indeed a better mode of knowing than the positivist mode, because it is truer to the fine texture of experience.* Is this true, and does Carter substantiate his claim? My own answer, in brief, is that it is a complementary way of knowing, and that Carter illustrates that well. His approach yields many valuable insights – but, as I have already suggested here, I cannot accept that one way displaces the other. All history, all science, all writing, is a process of selection – selection made in two ways, the obvious one of putting some things in and leaving some things out, and the less obvious but more basic one imposed by the structure of the discourse in which one is working.

By 'literature as a way of knowing' Carter clearly means the literary intelligence, that is, that of the writer and the literary critic conjointly, and he illustrates the workings of the literary intelligence again and again, above all in revealing the biases of selection imposed by historical discourse. What, for example, constitutes a primary source or 'document' for historians? Words. But whose words? The words of those who had the education and leisure to write; and in the early days of the settlement at Port Jackson this was a very small group, all, of course, drawn from the Establishment. Watkin Tench was one of the few early chroniclers, and he has been quoted *ad nauseam* by Australian historians.

43

So Carter goes to work on Watkin Tench, who was critical of the convict use of thieves' cant or 'flash language':

> Underworld cant – it is also the language of the roads, the language of illicit exchange, where objects change names as they pass from hand to hand. It is a language designed to baffle detection – and, significantly, Tench's first argument against it is the difficulty it creates in courts of law. Significantly, too, Tench sees the practice of this language as an integral part of the convict's physical practice: cant and crime, word and deed, cannot be separated. Predictably, Tench is of the opinion that an abolition of this unnatural jargon would open the path to reformation. Here we see the Enlightenment project of reducing the world to uniformity, replacing local difference with universal intelligibility. The path to reformation, like the road to Botany Bay, leads to prison. (pp. 317–18)

The multiple roads to Botany Bay

The reference to the 'road to Botany Bay' needs explication, since it is Carter's title, one that he uses to carry a range of meanings. Beginning with the obvious one, the road to Botany Bay is the path to the exploration and settlement of Australia. Then, at a more conceptual level, that road must be pushed back into the world view of the Enlightenment, which established ways of proceeding. But there is a third sense, the one referred to here. Once the colony had moved from Botany Bay to Sydney Cove, the road back to Botany Bay became something like the freedom road for southern slaves in the USA; it was seen by the convicts, and feared by the authorities, as a way of escape. That the attempts at escape – for example, to join the French ships anchored there for a time – had little success was less important than the idea of a road, the idea of escape that it kept alive. For the same reason, it was not important, indeed a complex irony, that there was no road physically. At best there was an Aboriginal pathway or a way through the woods. For Carter is linking two themes: the underworld language of the convicts, from which the Establishment was locked out, and the way through the woods as an anti-road, also excluding authority. Yet, says Carter, neither the convict cant nor their road constitutes a document or settlement, and thus they have no voice! Only Watkin Tench has a voice. In reconstructing the idea of that 'road', Carter is giving the convicts, and even the Aborigines, their voice. He does it with great ingenuity:

> In this sense, the convicts' recidivist habit of, in Collins's words, 'flying from labour into the woods', like their habit of retreating into an impenetrable argot, only served to bring home the urgency of clearing a space, of marking out conceptual, as well as physical, boundaries. For to call the road to Botany Bay a 'road' may have been a rhetorical means of suggesting the deceptiveness of appearances, but it also recognized the wood as the place of masked schemes, the place of highwaymen, unseen violence and strange translations (from white to black, from confinement to freedom), an environment predicated on the formalized and continuous transgression of fixed boundaries.
>
> So there is pathos in Tench's 'road' as well as irony: for while the convicts might prey on roads, in their own travelling they avoided them. Roads were for other people, for people who had an official destination in mind and were where they wanted to be at once. The more inconvenient they were for travellers, the better they suited the man of the road. For convicts, roads were proof that the fruits of travel were there for the taking, that roads bearing bullion distributed largesse. Roads were like laws, for

crossing at night. But confined to the high way of reason, how could Tench have acknowledged this? Only by settling it with roads of his own could Tench order Botany Bay's thieving spaces and deceptive meadows. (p. 319)

Carter and 'the literary intelligence'

Carter is exhibiting here one aspect of the literary intelligence, a sensibility to the shaping power of the written word. He reminds us, by quoting Adorno, that 'History does not merely touch on language, but takes place in it' (p. 325). History excludes all that is not quoted or written down, and thus it both reveals and *hides* our origins from us.

A second skill of the literary intelligence that Carter exhibits is the talent for the felicitous phrase. One example must serve. In describing the layout of Melbourne and most of our other towns and cities, we are told that 'Oriented towards the cardinal points of the compass, its grid was a container for real estate; its streets were conduits for auctioneers' (p. 204) (see colour illustration following p. 78). Evocative, highly suggestive of boom-town Melbourne – but the phrase also has the defect of such rich language. The meanings are not controlled or discriminated. Melbourne had wide, rectilinear streets, and it was a paradise for auctioneers – but is there a necessary connection? Prim and non-conformist Adelaide had equally wide, rectilinear streets. The plans of Melbourne and Adelaide have much in common when stripped of later overlays. But then so do Limerick in Ireland, and Delhi in India, all plantation cities imposed by the British. The initial function of the wide streets and grid favoured by colonial surveyors was that they made military control easier, and were well ventilated, at a time when disease was thought to be transferred by noxious vapours.

The third skill of the literary intelligence displayed by Carter, one we have noted already, is his flair for critical analysis. Some of his analyses are brilliant, while others succumb to the dangers of the method in being over-ingenious. Much of his probing uses dialectical pairs, especially the light and the dark, and the horizontal and the vertical.

> The horizontal and the vertical, the light and the dark: these are the twinned dialectical properties that characterize the burning light. They are also fundamental modalities characterizing the intimate space of the first white Australian travellers and settlers. (p. 264)

He fuses them by seeing pine trees as a dark flame:

> The pine trees marked a place of habitation. They stood at the focus of the road. They also stood for the inmates' verticality. They were like names, green, aspiring, vital. They stood for stability, endurance. They externalised a growing depth of attachment and perhaps their dark, bosky plumage assisted in suggesting their rootedness and antiquity. (p. 272)

This image of the pines (including Norfolk Island pines and the Bunya-bunya pines, the indigenous *Araucaria* so commonly planted beside nineteenth-century homesteads) makes them an extension of the candle in the window, an observation that attempts to find the meaning of a very widespread item of environmental behaviour in our past. He also comments in passing on the profound change in our imaginative life that must have followed the replacement of the flickering flame of

45

candle and lantern with the even and uniform light so easily there at the touch of a switch:

> moonlight and forest make visible to the traveller his own mythical origins, the jumble of gothic images that constitute his history. With the phantasmagoric clarity of a dream, they conjure up before him his own gods and keepers on the way. (p. 266)

The tension between horizontal and vertical is explored in many ways, peculiarly apt in Australia, where Carter can talk of vertical deprivation and vertical longing, both among explorers looking for a vantage point and the dwellers in flat suburbia, where vertical deprivation may be acute.

The best-sustained passage is part of an analysis of Matthew Flinders' journal recording his years on Mauritius (where he was held prisoner by the French). In 1805 he was given permission to lodge with a French family in the interior of the island. In the course of a journey through the mountains, he came to one of the island's spectacular waterfalls, and that gave rise to the following reflection:

> It appears to me that originally there had been only one great cascade or declivity at the mouth of the valley, but that the water draining through the crevices of the rock above caused pieces to fall down, forming another cascade. The same thing, happening further and further back in the course of time, has brought them to what we now find them; and it is still going on . . . thus nature proceeds in reducing all things to a level as well in the moral as the physical world . . . From reflections of this sort which I pursued much farther, I passed to the vicissitudes of my own life. I was born in the fens of Lincolnshire, where a hill is not to be seen for many miles, at a distance from the sea, and my family unconnected with the sea affairs, or any kind of enterprize or ambition. After many incidents of fortune and adventure, I found myself a commander in the Royal Navy, having been charged with an arduous expedition of discovery; have visited a great variety of countries, made three times the tour of the world; find my name known in more kingdoms than where I was born, with some degree of credit, and this moment a prisoner in a mountainous island in the Indian Ocean. (Carter, p. 196)

Carter's comment follows:

> It would be hard to imagine an autobiographical account that demonstrated more clearly how 'the centres of our fate', to borrow Gaston Bachelard's formulation, are located in 'the spaces of our intimacy'. For the profound tension between the horizontal and the vertical, which Flinders describes here, is not only a spatial opposition that goes back to his earliest memories of the Fens: it also expresses the ambiguity of Flinders's fate – where the ambition 'to get on' has led, despite himself, not only to horizontal, but also to vertical, advancement. But we notice that it is in the opposite, the threateningly vertical, environment of Mauritius that this process of reflection occurs. For much as the contemplation of the waterfall depresses Flinders, it also elates him. And this dialectical tension informs his recollection of home. If, in one mood, the flatness of the Fens, their open horizons, induces a delicious sense of retirement and refuge, in another, their unadventurous equilibrium provokes panic and revolt.
>
> We begin to feel both the horror of the plains and their attraction, a tension that, perhaps, only the map-making mind can bring under control. For, if the horizontal is the realm of Flinders's childhood, unenclosed by hills, open in all directions, the place where all futures are possible, then it is also the focus of annihilation, the inescapable point of return, where all variety is reduced to a level. In this latter aspect, it signifies the waste of vital energies, the dissipation of promise, the oblivion of an imprisonment

without walls. This is the double aspect of the plain: that it releases into nothingness. Directionless and equal, it inhibits motion, it resists exploration. The solution to the plain is the mountain. But this, too, has its ambiguous aspect. If, in the form of cascades, it encourages the current of associative thought, it also leads to the prospect of extinction. Obeying the vertical impulse of its nature, the cascade inexorably wears itself out, loses its nobility, in the vicissitudes of time grows perhaps serpentine. The hill Flinders once desired, perhaps, bounding the fen, a point of departure, grows, in adulthood, into a coronet of mountains, symbolizing not release but enclosure. The variety of the world, fanned out by his arduous career, is closed up again in an island as round as the first horizon. (p. 197)

This works brilliantly as literary analysis, deepening our awareness both of Flinders' use of language and of the quality of his experience, but I submit that it works so well because Flinders is himself exploiting a conventional literary trope, that of dialectical contrast, and that this coincides with Carter's dialectical preferences.

This is in contrast with passages in which the analysis does not work, such as the following, in which A. E. Howitt records a conversation with a shepherd near Bendigo, one of many whom the isolation of the bush had driven near to insanity:

I asked him what books he had read. He said he had read Burns's Poems and the Bible. And had read some of Shakespeare. Two of Shakespeare's dramas that he had read were Hamilton and Macbeth. These were all he could remember. But he sometimes got a map to read, which was very amusing. To trace out the roads, and the places of market-towns was very interesting. He complained much of the want of fruit in this country, and seemed to remember with a wonderful relish russet apples. 'Oh! those russets! They are beautiful fruit. I remember eating them somewhere in England; I don't remember where, but they were beautiful'. (Carter, p. 229)

Carter's analysis follows:

It is tempting to see in this poor man's obsession with maps a desire to find himself, to trace his own place in the world. As for the 'russets', so much bigger than the circles distinguishing 'market-towns' on the map, perhaps they represented the relish of roads, which, for all their strict linearity, implied the location of places, the roundedness of return. In the imaginary spaces of the map, he glimpsed perhaps a lost home; in the familiar names, a lost society. (p. 229)

Some temptations are better withstood. That the maps represent a desire to find himself is at least plausible, although simple homesickness would be my diagnosis. But in missing the apples, surely the poor man was doing no more than missing the apples, just as Australian housewives in America claim to miss Vegemite.

So I conclude that there is more than one way of knowing, with different claims on our attention. Paul Carter has made such claims, in this rich plum pudding of a book. It is now in paperback, and you should buy it, both for itself, and to encourage Faber & Faber to publish more such intellectual explorations.

Peter Fuller tells us that Carter is proposing two more books: 'The second will deal with migrations: the third, centred on Venice, will seek to apply spatial history to European cities' (Fuller, 1988). I hope he writes them, and I look forward to reading them, although I also share David Dolan's hope that 'if Carter and his spatial history are to fulfil their potential, they need a tough, sceptical editor for the next book' (Dolan, 1988).

Postscript

Carter, now at the Centre for Australian Studies at the University of Melbourne, is still writing, still for Faber & Faber. His latest book, *The Lie of the Land*, uses a title that I had considered for this book; it is a study of three men, T. G. Strehlow, the anthropologist; Giorgione (Big George), the Venetian painter; and Colonel Light, the surveyor and founder of Adelaide. Carter is still provoking the reviewers. Literary debates such as I have reported here offer an excellent snapshot of the current state of a culture, as have 'Demidenkismo', as my Italian academic friends call it, and Helen Garner's *The First Stone*. I see the vigour of the debates as a sign of cultural health. We have too few. The French have them all the time.

References

Adorno, T., and Horkheimer, M., 1973, *The Dialectic of the Enlightenment*, trans. J. Cummings, London: Allen Lane.

Carter, Paul, 1987, *The Road to Botany Bay: An Essay in Spatial History*, London: Faber & Faber.

Clark, C. M. H., 1987, 'A Historian Is Like a Cook Peeling an Onion', *Sydney Morning Herald*, 12 December.

Dening, Greg, 1980, *Islands and Beaches: Discourse on a Silent Land, Marquesas 1774–1880*, Melbourne University Press.

Dolan, David, 1988, 'New Territory, Faltering Steps in "Spatial" History', *Canberra Times*, 23 January.

Elliot, Brian, 1967, *The Landscape of Australian Poetry*, Melbourne: Cheshire.

Fuller, Peter, 1988, 'A New Road to Our History?', *Canberra Times*, 3 February.

Hill, Barry, 1987, 'Fresh Sign Posts in the Journey of Australian History', *The Times on Sunday* (Sydney), 13 December.

Howitt, A. E., 1971, *Come Wind, Come Weather: A Biography of Alfred Howitt*, see Mary Howitt Walker, Letter XX, p. 360, Melbourne University Press.

James, Clive, 1988, 'Bullshit and Beyond: The Road to Botany Bay', *The London Review of Books*, 18 February.

Lansbury, C., 1970, *Arcady in Australia*, Melbourne University Press.

Macintyre, Stuart, 1988, 'Demanding Delight', *Australian Society*, February.

Murray-Smith, Stephen, 1988, 'Naming time Down Under', *Age* (Melbourne), 16 January.

Porter, Peter, 1987, 'A Local Habitation and a Name: The Road to Botany Bay', *Times Literary Supplement* (London), 27 November.

Robinson, G. A., 1977, 'Journals', *Records of the Victorian Archaeological Survey*, no. 5, Melbourne: Ministry of Conservation.

Seddon, G., 1981, 'Eurocentrism and Australian Science: Some Examples', *Search*, no. 12, December.

Swanson, Elizabeth, 1987, 'Foreword', *Australian* (Sydney), 17 December.

Thompson, Christina, 1987, 'Botany Bay, from Spaces to Places', *Herald* (Melbourne), 15 December.

Ward, Russel, 1987, 'Getting Lost in Space', *Weekend Australian* (Sydney), 5 December.

5 A Snowy River reader*

Writing environmental history

I spent five or six years exploring the Snowy River and studied river and records intensively in almost every spare moment for the last three of them. I have come to love the Snowy, and want to share that love by writing a book about it, but have found it hard going.

There are three problems. The first is that although the Snowy is in the most densely populated south-eastern corner of Australia, its immediate setting is wild, rugged and inaccessible, and it has been little studied, so there are uncertainties and great gaps in the available information; it is not possible to give a definitive account of any aspect. Despite this, I have gathered a wealth of information about the river and its natural and human history, but that leads me to my second and third problems: what to put in and what to leave out; and how to organise my material. These problems illustrate some general problems in writing environmental history; in fact they also forced me to attempt to define 'environmental history', and to ask myself how it differs from other kinds of history, and why anyone should want to write it or read it. When I began I thought that the answers to these three conceptual questions were self-evident, but perhaps they are not. If environmental history – by which I understand the reciprocal interaction between man and his setting, or, more expansively, how we have changed a given environment through time, and how it has changed us – is as significant as I think it to be, then why isn't there lots of it, with a well-established tradition for me to follow? I shall return to this lofty question after I have explored further my three primary questions about the Snowy.

The areas of ignorance or uncertainty are many. For example, little is known about the Aborigines of the region, and Josephine Flood's study *The Mothhunters* (1980) was a pioneer work. It has been followed by several further studies, but the information base is necessarily meagre, restricted to the partial accounts of early observers and to the gleanings from the excavation of campsites. What we do know

*First published as 'A Snowy River Reader', *Meanjin*, vol. 45, no. 3, 1986, pp. 309–30.

is that they were there, making a living along the whole length of the river, most of it now uninhabited.

Other matters are the subject of scholarly debate: interpretations of the geomorphology, the nature and timing of the so-called 'Kosciuskan Uplift', the discovery of Kosciusko itself, the evolution and recent history of the Bogong moth migrations; the role of fire, both natural and man-made, in the distribution of vegetation. Debate and controversy are a part of all scholarship, but the topics I have listed above have generated two opposing camps with little prospect of an easy truce, and the substance of the debates is not about values and points of view. It is more elementary, concerned with the facts and their interpretation.

On to the second of my questions: criteria of inclusion and exclusion are derived from central organising themes and by decisions about the intended 'audience' for a book. My major problem is the structural one and that alone makes it hard to decide what to put in and what to leave out, but there are subsidiary problems as well. For example, there is the question of privacy. The families who have lived along the Snowy are relatively few in number, and they all know each other. One of the great-grandmothers of one family I describe was a convict, and there are others I know of. Most younger members of such families now accept their past calmly, but not all do, and one can see why, given the very firm social and economic stratification in the region (as in much of rural Australia). The big land-owning families, begun by those who came as free men, with some capital, have often established dynasties. They do not form a squirearchy and are rarely wealthy, but they generally have a fair level of education, usually from one of the Sydney schools, which gives them a metropolitan base, and many of them have travelled outside Australia. The families of the convicts and labourers of the early days have sometimes moved up or moved out, but others are still to be found as labourers, small landowners, storekeepers and tradespeople (I have found some by direct enquiry, but I know of no detailed demographic study that goes beyond the work of Hancock (1972, pp. 82, 112 and *passim*)). Some of them are sensitive about their origins, so I have not recorded them by name. There are also personal and social attributes that are worth recording, but again one must tread very warily. Some points can be made only in a general way: for example, many of the real bushmen are overweight. The old-timers like Bill Wroe and Whit Ingram are still slim and wiry, but they also still ride: straight as a ramrod on a horse, bow-legged and a little stooped on the ground. But nobody has ever walked much in this horse country, and now that the middle-aged have traded in their horses for a Land Rover, they drive *everywhere*, and their figure betrays it. Male and female domains are usually very sharply defined. Muddy boots come off at the back door, and stockinged feet pad the house. The men wash up in the wash-house. The bathroom is for women, as the decor shows clearly. But it would be invasive to pursue these details, and more so to attribute them to individuals.

Another major problem of selection is that of detail, and this is a problem at several levels. It is easy to use all the place names when writing of an area that is well known, but hardly anyone knows all the place names along and near the river, and, to make matters worse, many places have several names: for example, the Pinch River is also the Moyangul and so on. I find myself drowning in the detail, not just of names, but of topography and geographic linkage – how one gets from A to B, for example. In trying to give a comprehensive view of the river from source to mouth and of the history of its use, one needs further to call on the earth sciences and biology. Again, there is a problem in deciding how much a potential reader can be assumed to know. What, in short, needs explaining, and what can be assumed?

Choosing an organising framework

But these problems are minor compared with the problems of organisation. Considering the alternatives is a way of examining some familiar thought patterns and their limitations, and the exercise has also given me an oblique perspective on Australian history and society. I shall discuss the major options below.

Geographic sequence, following the river from source to mouth. The river itself is a geographic entity, and such a sequence would be appropriate to a travel book, but it is an awkward way of managing themes and history. The river is not a geological or biological unit; neither is it a historical, social or political unit. The Nile, the Rhine, Ganges, Mississippi all have a historical, social, economic, and political significance. The only river in Australia of which that can be said is the Murray. Australia is organised around ports, roads and rail, not along rivers. That is in fact one of the things that I have 'discovered' about the Snowy. It is *there*, a magnificent natural force. It has mythic stature, through one single poem. But it has no historical, social or political reality; *parts* of it have mattered a great deal in various ways at various times, but only as a part of a different fabric. The most significant fact about the Snowy is that our society has never really known what to do with it. There is not even an accepted name for the area that I am discussing. The 'Murray Valley' and the 'Hunter Valley' are well known, but one cannot speak of the Snowy Valley – that is too gentle a word, a prosperous, fertile, smiling word, for the narrow, tortuous cleft in the rocks through which the Snowy hurtles to the sea. The 'Snowy catchment' would do, but might be taken to refer to the upper part of the river only. The 'Snowy Drainage Basin' is correct, but too formal for popular use, and in any case it is not much like a basin. There are names only for parts – the Monaro, the Byadbo, the Wulgulmerang and the Gelantipy plateaux and the Orbost flats, of which only the first and last are well known, and all five cover only half the river. It has no regional name because it has no regional function for our economy or society.

Historic sequence, a chronology of events or their interpretation, offers a conventional organising principle, but, as noted above, the Snowy River per se does not really have a history of its own. The lines of history do not run that way, although the events that have taken place there can be seen as part of a broader sequence, as we shall see.

Natural history is a thematic sequence geographically applied: the geology, soils, flora and fauna are discussed in turn, with the opportunity to indicate interrelationships. This is a standard scientific approach and it would work, although for a limited audience. The natural history of the Snowy has not been assembled to date, and such a compendium would undoubtedly have value. It would be arbitrary in that the geological, pedological and biological boundaries do not coincide with those of the river catchment, but the river is a natural hydrologic or geomorphic unit, and it is a major biological corridor. The river traverses greater ecological diversity than any other river in Australia, from alpine, montane, sub-arid grassland, sub-arid sclerophyll savannah to dense rainforest, and that too is of biological interest (see colour illustration following p. 78).

This principle of organisation would be logical enough, but rather dull, because natural history excludes people, whereas people are central to my story. Natural history forms a necessary part of a history of land-use, that is, of one half of the interaction between our species and the environment that I conceive as 'environmental history'.

History of land-use requires a compound of organising principles, a chronology of events and a set of themes. This is a major concern, and a broad outline of such a sequence can be given easily, as is set out below.

There seem to be four or five broad categories of land-use: that of the hunter-gatherers; that of semi-nomadic pastoralists; changing slowly into a settled pastoral economy, with service towns, roads, railroads and other services; and that of the urban industrial society, which thoughtlessly murdered the river and now uses the corpse for recreation (perhaps this necrophiliac use is a fifth category). Much of the information about each of these phases is inherently unreliable, because each has been the subject of myth.

The Aborigines have been 'observed' and presented as Noble Savages, primitive brutes, or children; and, more recently, as Economic Man, driven to migrate hundreds of kilometres up the Snowy by the need for the supposedly protein-rich Bogong moth; or, alternatively, as Ecological Man, living in conscious harmony with the natural environment, and a Lesson To Us All.

The pastoral phase became a Heroic Age with Banjo Paterson and the *Bulletin*, although this was and is an urban construct, reflected more recently in, for example, the advertising for Marlboro cigarettes. Accuracy has never been important to this myth: The Man from Snowy River ads for Marlboro were mostly filmed in the South Island of New Zealand, which photographs better. Banjo Paterson also got much of his detail wrong. The real cattlemen from Snowy River were mostly illiterate, exploitative and very destructive. Large tracts of productive land were rendered useless within a few decades. Many of the men were, however, magnificent horsemen and they and their women-folk were remarkably resourceful to survive and reproduce in some of the most spectacularly broken and difficult terrain in Australia.

The Snowy Mountains scheme ran the river west through the Divide for hydropower and irrigation in the Murrumbidgee and Murray valleys, now beset with major ecological and economic problems. The project was never properly costed, and it has been argued that it was far from cost-effective. It was also dubiously legal at the time; if challenged, it would have had to have been defended by invoking the powers of the *Defence Act*. It had some social benefit in providing employment for 'DPs' (displaced persons, many of them from the Baltic States), who were *required* to work in rural areas for a time because of the current decentralisation policy, at a time when rapid mechanisation was dramatically reducing the demand for farm labour. This 'melting pot' function, however, was vigorously promoted by the Snowy Mountains Authority and has become part of the myth. The death rate was high. The engineering and project management was generally of a very high standard, and it was undoubtedly of psychic value for post-war

Australians in seeing themselves as capable of great peacetime deeds at the national level, and it was thus promoted by the Department of National Development. The scheme can therefore perhaps be seen as a sacrifice on the altar of national identity, like the other great human sacrifice at Gallipoli. We live by myth.

There was, of course, no environmental assessment whatsoever of the consequences for either the Murray or Murrumbidgee river system. The benefits of irrigation were taken to be self-evident at the time, and the Authority was able to boast freely that the additional irrigation water made available by the scheme was 'free' – that is, paid for by the metropolitan consumer of electric power. It is now generally agreed that one reason why irrigation water has been so thoughtlessly used in the Murray Valley is that it has been too cheap. The Murray–Murrumbidgee are now deeply in trouble. The beheaded Snowy is dead from Jindabyne to the Quidong, where the Bombala and Delegate rivers form its functional headwaters today. Above the Quidong, you can step across the mighty Snowy, a cot-sized trickle in a king-sized bed. This is thus the history of a throwaway river.

Such an approach, that of land-use history, looks like an adequate framework at first sight, but problems persist. The first and most obvious is that each of these phases belongs to different segments of Australian history, not just in time, but in their linkages; moreover, they are, at first sight, not significantly linked to each other except by the accident of geography, and partly for this reason, each requires different background information. However, that each of these phases seems discrete, with little to link one phase with the next, is true only if one looks for the links within the Snowy region. That is a problem in writing most regional history – the major events are generated outside the region. The phases that I have listed are linked if we consider them as successive transformations of Australian history, itself a colonial response to 'four long waves of expansion and decline in the capitalist centre' (Mandel, 1975; see also Berry, 1984). Aboriginal land-use remained discrete, until it began to be displaced by the pastoral economy, which was succeeded by mercantile capitalism, followed by industrial capitalism and by corporate capitalism.

Three of these phases are exemplified very positively in the Snowy. The story of its European settlement is the story of pastoral expansion. The Snowy Mountains scheme is very much a product of industrial capitalism and of the values which upheld it. Without those values it is virtually incomprehensible, with them it seems necessary, almost inevitable. The post-industrial society, that of corporate capitalism, brought with it a new set of values; a quest for legitimation brought with it a concern, among others, for the environment, and hence support for national parks, while it also made available the leisure to enjoy them. With this set of values, it seems unthinkable that the Snowy should be seen other than as a great natural resource, to be jealously guarded as a national heritage. Within a decade, the river changed from a mere water supply, to be manipulated at will, with no inherent value of its own, to its current status – still largely unknown and unmanaged, but replete with heritage values.

It says something for the flexibility of our sense of national identity that many people seem capable of maintaining both sets of values simultaneously regarding the Snowy: they can be proud of our great engineering achievement, and be thrilled by the grandeur of a wild river. That these values are not seen to be incompatible may be due in part to an uneasy feeling that we have little to be proud of, and that we should therefore lose no opportunity. Major industrialisation came late to Australia, and it has had such a short and recent history that it may be hard to accept its demise, and in any case we have had few major engineering achievements other than the Sydney Harbour Bridge and the Opera House, which can be regarded as sources of national pride by non-Sydneysiders only when they are overseas. The change in values from industrial capitalism to corporate capitalism is not to be traced within the Snowy region, and does not form a part of this essay, but it is certainly drama-tised by the events recounted here, in that its consequences have the appearance of abrupt discontinuity at this regional scale.

The phase of mercantile capitalism is least dramatically represented in our region, or rather, it is more often represented negatively, by the things that didn't happen rather than by the things that did. The major events were the consolidation of the pastoral industry in the Monaro, by land purchase, the construction of fences, dams, substantial homesteads, roads and service townships. This story has been told well by Hancock (1972) in *Discovering Monaro*. In the 1890s, Cooma was linked to Sydney by rail, by way of Goulburn. The other area of settlement was the Orbost flats, but this was several decades later than the Monaro. Orbost was linked by rail to Melbourne only in 1916 – or almost linked, as the rail stopped at Newmerella on the west side of the Snowy, where it stops today more than eighty years later. Thus both Cooma and Orbost were at the end of the line, and this symbolised their status as far-flung outposts of metropolitan dominance. All the significant investment decisions were made in Melbourne and Sydney, and these determined land-use, within the limits of its natural capability. It is tempting to see what in fact has hap-pened – which is very little – as almost inevitable, given the remoteness of the area, but that is not so. The region might have had a very different history if different decisions had been taken outside it.

One national policy decision to affect the Monaro and the Snowy was the ultimate choice in 1909 of Canberra as the site of the national capital, overturning an earlier decision in 1904 to build it near Dalgety on the Snowy River. The 1904 Act, opposed by New South Wales, was therefore never put into effect; if it had been, the environmental history of the region would have been very different. Canberra has been a poor site for a national capital because it has no natural place in a regional economy of a thinly populated grazing landscape, although of course it has had a great impact on the region. A similar city in the Monaro would have had a different kind of regional significance. The waters of the Snowy would doubtless have been used to generate hydropower, but the water would have remained in the Snowy. The major effect would have been to improve communications with – doubtless – both rail and highways to Eden and its fine harbour on Twofold Bay, and to Orbost in Victoria – thus to the growth of intensive agriculture in the coastal valleys, and of much more substantial development of the south-east coast of Australia.

It seems then that I have found a way of ordering my material into a history of land-use, by looking for the links elsewhere. But that would be only half the story, 'a study of man's impact on his environment' (the subtitle of *Discovering Monaro*). But what of the interaction? *Why* have we behaved in particular ways at different times? Such a question can be answered at many levels, and it can be broken down into smaller questions: for example, who liked the country enough to stay there in the first half-century? The answer: the Scots, who are substantially over-represented. It is easy to speculate why. It is also easy enough to see why the rightness and greatness of the Snowy Mountains scheme seemed beyond question in the 1950s. In fact I have already introduced this theme, the impact of the environment on us, by using the word 'myth'. To use such a word is to signal concerns that go beyond social, political and economic history into the domain of cultural history. To understand the variety of behavioural responses to the Snowy River catchment we need to look at the ways in which it has been perceived.

Perceptual history is then a fifth option, offering a sequence of records of the way in which the river has been seen, by explorers, photographers, artists, natural historians, cattlemen, miners, engineers, hydrologists, bushwalkers . . . There have been almost as many rivers as there have been observers, and that is in the end why the river is 'incoherent'. There can be no single view of it.

One of the more interesting is that of the Reverend W. B. Clarke, who published an account of his geological explorations in 1860: *Research into the Southern Goldfields of New South Wales*. Clarke was a good economic geologist, and it is obvious that he much preferred field exploration to looking after his parish on the outskirts of Sydney town; but he feels obliged to justify his tastes, and therefore breaks into the rhetoric of the 'sublime':

> The most gorgeous cathedral, filled with holiday worshippers, is not more pleasing in my recollection than that noble landscape by which we were surrounded, and the company of pilgrims who stood by me under a burning sun on the side of the hill, listening to my homely words of encouragement and exhortation. It is to be hoped that I may never be reproached with forsaking my calling to seek for the gold that perisheth, for the judgements of the Lord, which I proclaimed amidst the mountains, 'are more to be desired than gold, yea, than much fine gold'. (p. 119)

Nevertheless, the fine gold was the object of his explorations and, purple passage over, he returns to professionally prosaic appraisal of the geology and the likely distribution of precious metals. Other passages combine the rhetoric of the picturesque with flat economic evaluation, for example:

> This part of the course of the Snowy River is wild in the extreme, and the ranges about Walgalamarang, and all through to the south-east, towards the Deleget country, is a collection of broken, steep and almost inaccessible masses. It has never been thoroughly explored, though the most considerable portion of it is north of the boundary. On another occasion I approached this defile of the Snowy River from the south-eastward; it was there equally inaccessible. The amount of fluviatile and transported surface drifts must be enormous; and doubtless they contain gold. (p. 123)

The 'picturesque' is also shaping rhetoric in a somewhat later account by W. H. Ferguson (1899), also a geologist, of the Snowy gorges across the Victorian border (see colour illustration following p. 78):

> The scenery is wild and rough and grand in the extreme. In no place else in Victoria are there such dizzy precipices, such sheer bluffs, or gorges with such vertical sides. In places the river is hemmed in between rocks which leave but a 30ft waterway. In others the waters of the stream ripple over the gravel beds with a width of 7 chains. In places the river is a wide still pool; in others it is a soaring rushing rapid, which plunges tumultuously over a bouldery 10ft drop . . .

In 1937, Arthur Hunt and Stanley Hanson made the first successful canoe trip down the Snowy all the way from Jindabyne to Marlo at its mouth. 'Don't be a fool, Arthur', advised one Monaro friend, 'there are places down there where even the blacks won't go'. But he went, took two months to make the trip, and described it for the *Sydney Mail* in three articles (Hunt, 1937). He also used the rhetoric of the picturesque, but it is a laconic bushman's version:

> We had the crows with us, and we always imagined they were the same crows right through the trip. When we pulled our canoe out of the water for the last time, it was in the darkness of a wild, rough night, and the scream of the wind sounded like their disappointed curses.
>
> Many times during the two months we were on the trip one of us would pause, shake his fist at the crows, and say: 'You haven't got us yet, you black brutes!'
>
> The Snowy River is a strange, lonely, and wonderful river, and breeds strange thoughts.
>
> Stan, however, livened up the proceedings a little by finding a brown snake – one of several we had seen that morning. He teased and tormented it, hitting it across the head with his hat when it struck at him.
>
> He got a forked stick, jammed the snake's head hard on the ground, picked it up and played with it. When he put it down, I put an end to his childish fun with a large rock.
>
> I think Stan was annoyed because I killed that snake. (21 July 1937, p. 51)

The harsh cry of the crows is the appropriate background music, and Stan's game with the snake is a smaller version of their game with the river, for which they show the grudging respect – almost love – that many Australians used to feel for a tough adversary, like Johnny Turk in World War I. After their first leg through the Beloka Gorge below Jindabyne, Hunt and Hanson spent a couple of nights at Dalgety, mostly in the pub:

> We would meet someone for the first time and be introduced as one of the coves who were going down the river. He would say: 'You got this far, did you? How did you get on coming through the gorge?' We would tell him in as few words as possible.
>
> Then would follow that long-drawn 'Ummmm!' which shows that an Australian is getting into gear to say something, and the little play would conclude: 'Yes, it's a bit rough up in there; but you wait till you get further down. You'll never get through. You'd better come and have a drink.' (21 July 1937, p. 51)

They did, but continued with their trip, and were soon into the very rough country between Ironmungy and the McLaughlin River.

> For the next four days we averaged less than a mile a day. We worked from daylight until it was too dark to see, very often going without our midday meal. It was a canoeist's nightmare. We were well behind time, and we began to have doubts that our flour and tobacco would last out until we reached Williams.
>
> We slipped and fell, dragged, hauled and sweated. We pulled our sleeping-bags on each night and were asleep before we had time to finish our after-supper smoke. We almost forgot what it felt like to be in the canoe; it was just portage, portage, portage through the roughest country it was possible to imagine.
>
> There were no pools – only small falls, bad rapids, rocks, foam, and the dull roar of the river. We were reluctant to do any shooting as we were afraid of rain, which would have meant abandoning the canoe, as a few more feet of water in those gorges would have left us no room even to climb along the banks.

The banks were masses of boulders, with the sides of the gorge rising sheer in some places for hundreds of feet. Our hands cracked and bled, Stan's back was getting worse, the canoe was leaking again – altogether we had a pretty rough time.

We found that in one place the river completely disappears and runs underground for fifty yards. Very few people on Monaro know of this, and many were inclined to doubt the authenticity of the tale.

The truth of it was brought home to us when we camped near the Stone Bridge, as it is called, and had to walk 100 yards upstream before we could climb down to the water's edge to fill our billies. The blacks used this natural crossing when they were travelling backwards and forwards from what is now Victoria. (28 July 1937, p. 15)

I have quoted Hunt at length because his words bring the river to life for me far more than the purple passages of Clarke, and his tone of voice is rare in written accounts. It is exactly the tone of men who have lived by and worked around the river, but very few of them have put pen to paper. Miners, engineers and hydrologists have discussed the river and its surroundings in more prosaic and utilitarian terms, but even the rhetoric of utility tends to break into hyberbole. This is particularly well illustrated by Robert Menzies in a speech delivered at the opening of the Tumut Pond dam in 1958:

> In a period in which we in Australia are still, I think, handicapped by parochialism, by a slight distrust of big ideas and of big people or of big enterprises . . . this scheme is teaching us and everybody in Australia to think in a big way, to be thankful for big things, to be proud of big enterprises and . . . to be thankful for big men. (quoted in Wigmore, 1968, p. 194).

Perhaps Menzies was correct in this assessment, in that, as I have suggested above, the scheme has probably had more psychological than utilitarian value, although it has to be said that Menzies was not one of 'the big men' when the scheme began under Chifley; indeed, it was he who questioned its legality. And of course the scheme was promoted on utilitarian grounds, even if they proved illusory, requiring a notable sacrifice of non-utilitarian values. The case for these non-utilitarian values has never been better put, again using the rhetoric of the sublime, than by B. U. Byles, the forester who made the first major soil conservation study in the upper catchment in what is now Kosciusko National Park. The occasion was the proposal by the Snowy Mountains Authority and its big men to dam Spencers Creek, a high-level headwater of the Snowy, a proposal strongly opposed by the Academy of Science, but abandoned by the Authority on pragmatic grounds that the glacial moraine was too unstable for secure foundations. Byles' words: '*all* the water in the high levels of the park must not be converted to power; *some* of it must be left on the altar of the gods'.

Banjo Paterson also wrote in the romantic-heroic mode. All the writing about the river is in this vein, using the rhetoric of the picturesque or the sublime. Either that, or it is utterly prosaic, the prose of hydrologists recording water flow or the like. There is almost nothing in between. Perhaps because our society has not been able to find much direct utility for the river, our imagination seems unable to encompass it. The record is one of environmental abuse and lofty prose. Now that what is left of the river is dedicated to a non-utilitarian use, that of national park, we may learn to see it in new ways.

57

The altar of the gods: this view of the summit of Bogong shows rather well that our 'alps' are a dissected peneplain. NEVILLE ROSENFELDT

Postscript

Part of my problem in writing environmental history stems from the linearity of language, where I wanted a polyphonic account. Of course I did not solve the problem, but I finished the book, *Searching for the Snowy: An Environmental History* (Allen & Unwin, 1994). In the end, I followed geographic sequence from source to mouth – in the end there was no real choice – but in segments, so that I could interpose chronological sequence for each segment (mostly history of settlement, history of land-use, and history of scientific exploration) along with natural history. I then tried to weave into the fabric samples of perceptual response to give richness and diversity, with some analysis of rhetorical styles, naming practices, such thematic control as I could manage, and a good deal of cross-referencing – like a juggler keeping half-a-dozen balls in the air. The metaphor of 'A fugue for six voices' attempts a similar polyphony in this book: I want all the six themes to be interactive, with a simultaneous force, but of course I have to lay them out one after the other. In 1996, my son and I made a CD ROM on the Snowy, using a technology that allows different opportunities for reflexive interplay.

I write a second postscript with hope: *Searching for the Snowy* is a book about a river dying in its upper reaches. This is at last a matter of public concern, and under review. Friends of the river in the Monaro are asking for an increase in the waters dammed at Jindabyne from a niggardly 2 per cent to at least 25 per cent. A significant release of water from the Hoover Dam in America to flush the Colorado gives us new heart.

References

Berry, M., 1984, 'The Political Economy of Australian Urbanisation', in D. Diamond and J. B. McLoughlin (eds), *Progress in Planning*, vol. 22, part 1, Oxford: Pergamon Press.

Clarke, Rev. W. B., 1860, *Research into the Southern Goldfields of New South Wales*, Melbourne: Macmillan.

Ferguson, W. H., 1899, 'Report on Geological Survey of Snowy River Valley', in *Geological Survey of Victoria Progress Report No. 11*.

Flood, J., 1980, *The Mothhunters: Aboriginal Prehistory of the Australian Alps*, Canberra: Australian Institute of Aboriginal Studies.

Hancock, W. K., 1972, *Discovering Monaro: A Study of Man's Impact on His Environment*, Cambridge University Press.

Hunt, A. L., 1937, 'Down the Snowy by Canoe', *Sydney Mail*, 21, 28 July, 4, 11 August.

Mandel, E., 1975, *Late Capitalism*, London: New Left Books.

Seddon, G. 1994, *Searching for the Snowy: An Environmental History*, Sydney: Allen & Unwin.

Wigmore, L., 1968, *Struggle for the Snowy: The Background of the Snowy Mountains Scheme*, Melbourne: Oxford University Press.

Perceiving: the eyes and the mind

If I were an ant, my environmental problem would be the anteater.

An African tribe was brought out of the dense jungle in which they had lived all their lives, in what used to be called French Equatorial Africa. When they came to the clearing, they at first tried in vain to shake hands or clasp the arms in greeting of people who were in fact many metres distant from them. In the dense jungle, if you could see other people at all, you were very close to them, and they therefore assumed that the figures they saw in the clearing were doll-sized people who were nonetheless very close to them.

This story was told by a famous French anthropologist, and it is supposed to be true. Maybe it is, maybe not, but it is a nice little story. The good thing about stories is that we can make  them mean whatever we like. The points that I would draw from this one are: that behaviour that is appropriate in one environment may not be appropriate in another (or, as George Bernard Shaw puts it, 'Do not do unto others as you would that they should do to you; they may have different change; and that even some of tastes'); that environments our most basic perceptions, such as judging distance, which we take to be an objective response to the real world, are in fact conditioned by our prior experience, which is not necessarily relevant, and may be positively misleading.

Differences in perception are explored in all these essays, which, as it happens, are in chronological sequence, from 1976 to 1990. The first, here being Chapter 6, The evolution of perceptual attitudes, was written to introduce a seminar I was able to mount as then chairman of the Australian Committee for Man and the Biosphere (a UNESCO program), and although it is now dated, it is a record of how things looked at the time. It then fell to my lot to edit the ensuing book, published as *Man and Landscape in Australia* (reprinted once, but long since out of print, and now something of a collectors' item, unusual for the products of the Australian Government Printing Service). It is a book that illustrates the theme of this section very well, with essays from hands as diverse as those of Jack Mundey, the hero of the day (he persuaded building and construction unionists to impose 'green bans' on projects that were seen at the time as environmentally unsound); the distinguished scientist Sir Otto Frankel; and a developer, Sir Paul Strasser, at the time the Joint Managing Director of Parkes Development Pty Ltd. All three provided thoughtful essays, and Frankel, himself a migrant, wrote a perceptive tribute to the indigenous vegetation: in Australia, he says,

> the contrast between the Australian and the exotic elements [in the vegetation] is peculiarly prominent, more so than in most other places; so much so, as I have determined in casual opinion polls, that most people are conscious of the personality split in the Australian landscape, and that they not only perceive and register it, but that it is a subject of interest and concern.

He then contrasts this with both Europe and New Zealand, where he worked for some years. 'In large parts of Europe, there is little left of the indigenous structure of any landscape . . . Few Englishmen or Germans would know whether a plant was native or introduced, nor would they care'. As for New Zealand, native and indigenous plants are for the most part segregated ecologically and geographically, with the indigenous ones confined to the 'unusable parts of the country'. Most New Zealanders live in a 'transformation landscape' with a flora that is alien to the country (Seddon, 1976, p. 49).

I added an epigraph to the book as follows:

> An ecological vision of Australia has been slow to develop, partly because her white settlers have come from an energetic, transforming race with a culture that has evolved in a very different environment. Australia might have been settled by Spaniards, or perhaps by an agricultural people five hundred years ago, using the genetic material to hand – grasses, seeds, fruits, marsupials – for centuries of selective breeding, parallel with that of Eurasia and the Americas, to evolve an indigenous agriculture. Prize kangaroos might have been kinder to the Australian landscape than the sheep and the rabbit, and would have fostered a different understanding of it. But this is fantasy. What happened was the superimposition of European practices, with varying degrees of adaptation, and a generally limited perception of natural ecosystems. The theme of the seminar is thus the interaction between imported cultural notions – economic, social, aesthetic – and response to the indigenous environment. It begins historically, and then turns to implications and the future, to ask what kind of Australian landscape we can have and want. (Seddon, 1976, p. v)

That we see what we have learned to see is illustrated in Chapter 6, by the comparison of early perceptions of Port Jackson, although from the beginning the perceptions depend on the observers and their needs. If it is true that we tend to see what we have learned to see, it is also true that we modify and extend what we have learned by seeing: if not, the scientific revolution could not have taken place. Yet scientists, too, may be hampered by preconceptions – see Chapter 7, Eurocentrism and Australian science. Chapter 8, Figures in the landscape, is an exploration of the relation between man and his environment as illustrated through time, from 1832 to 1987, from a single vantage point, the view of the site, town and city of Perth from Mt Eliza. From the beginning of European settlement, Perth below was sketched, painted and photographed, and still is, from the cliffs above it, and this incredibly rich record would permit a much fuller account of the evolution of perceptual attitudes than is given here, although the theme is pursued more fully in the book in which this essay became a chapter, *A City and Its Setting* (Seddon and Ravine, 1986). The title of this chapter indicates its focus: the changing role of human figures in drawings and photographs until the 1930s, when they disappear. This indicates a change in the way the view is perceived as human setting, a change that is both remarkable and little studied.

Chapter 9, Dreaming up a rainforest, is an attempt to analyse several different types of discourse about rainforest, to see what we can learn from each, and what we bring with us in our perceptions. We bring the bright enchanted forest of the Douanier Rousseau, the Garden of Eden, the serpent, and Conrad's *Heart of Darkness*. The enchantment and the heart of darkness are in us and not in the rainforest: or, as Henry David Thoreau puts it:

> It is vain to dream of a wildness distant from ourselves. There is none such. It is the bog in our brain and bowels, the primitive vigor of Nature in us, that inspires that dream. (in Schama, 1991, p. 578)

Chapter 10, Home thoughts from abroad, is an essay in perspective. The theme, that what a place looks like depends on where you are looking from, is banal – but Australia can offer some striking exemplifications.

References
Schama, Simon, 1991, *Landscape and Memory*, London: Harper Collins.
Seddon, G., 1976, 'The Evolution of Perceptual Attitudes', in G. Seddon and M. Davis (eds), *Man and Landscape in Australia: Towards an Ecological Vision*, Australian UNESCO Committee for Man and the Biosphere, pub. no. 2, Canberra: Australian Government Publishing Service.
Seddon, George, and Ravine, David, 1986, *A City and Its Setting: Images of Perth, Western Australia*, Fremantle Arts Centre Press.

The evolution of perceptual attitudes*

The common aim of the essays in *Man and Landscape in Australia* is to show the evolution of the perceptual attitudes of Australians to their land. Some are descriptive: that is, they attempt to show how Australians have in fact perceived their environment through time (and for this reason, much of the book is historical). Some are also prescriptive, suggesting ways in which our environmental perceptions might be heightened and more finely attuned to the needs of the land that is now home. The *Programme on Man and the Biosphere*, which occasioned this book, is biologically oriented, and the subtitle, *Towards an Ecological Vision*, derives from that orientation. All contributors to the book would agree with the general proposition that Australians should become more sensitive to the special needs of their unique landscapes, but agreement probably ends there. We have different prescriptions, and different diagnoses of the past, some of them primarily cultural, some economic (for example, Bruce Davidson shows very clearly the major economic constraints that have determined the broad patterns of Australian agriculture, and suggests in doing so that they could hardly have been different from what they were).

It is not surprising that there are differences of opinion. It cannot be assumed that we know a great deal either about what Australians 'see' or about the forces that have shaped their perceptual worlds. Further, there is no such thing as 'the Australian environment': Balmain is not much like Eucla. In 1976, thirteen million different people saw a great variety of different places, in different ways, at different times, and any one person sees the same place in many different ways in the course of a day. For example, in 1971 Melbourne had 108,000 people of Greek origin, most of them living in and around South Melbourne and Brunswick, and thus living in cities within a city. There have been few studies of their perceptual responses to their environment. Contemporary perceptual studies of any kind are few (see Walmsley and Day, 1972). We have been a little better served by historical studies,

*First published as 'The Evolution of Perceptual Attitudes', in G. Seddon and M. Davis (eds), *Man and Landscape in Australia: Towards an Ecological Vision*, 1976, Canberra: Australian Government Publishing Service.

especially by Bernard Smith's subtle and provocative *European Vision and the South Pacific* (1960), Geoffrey Blainey's *The Tyranny of Distance* (1966) and studies such as R. L. Heathcote's 'Drought in Australia: A Problem of Perception' (1969). Sir Keith Hancock's history of land-use, *Discovering Monaro* (1972), may mark a new beginning among professional historians, who have generally ignored the land.

Part of the difficulty with such studies is that it is not always clear what should count as evidence. The only perceptions we know directly are our own. We guess at the perception of others from what they say they perceived, what they record, what they omit, how they behave. Poetry, painting, old films, advertisements and newspapers are all records, but they all need interpretation. Early pictorial artists drew the indigenous flora and fauna, the topography and the Aborigines in certain ways, partly because of current traditions of draughtsmanship, and also because of the function of their work (for example, to serve as scientific illustration rather than decorative wall hangings), partly because of individual idiosyncrasy, or limitations of skill with the pen, partly because of cultural preconditioning (which led, for example, to the portrayal of Aborigines first as Noble Savages, and a few years later, in reaction, as comic scarecrows), and partly as a genuine visual response to the object before them. This last component, one which some would call the only genuinely perceptual component, is not in fact separable from the cultural and biological predisposition to see in certain ways.

If it is hard to interpret individual perceptions, it is harder to make useful statements about those of a community. From the outset, individual responses to the land varied widely, as the following early descriptions of the site of Sydney show well enough. The first is from Major Robert Ross:

> I do not scruple to pronounce that in the whole world there is not a worse country than what we have yet seen of this. All that is contiguous to us is so very barren and forbidding that it may with truth be said here that Nature is reversed; and if not so, she is nearly worn out . . .

The second is from Lieutenant Ralph Clark:

> This is the poorest country in the world – over-run with large trees, not one acre of clear ground to be seen . . .

The third is from Captain Watkin Tench:

> The general face of the country is certainly pleasing, being diversified with gentle ascents, and little winding vallies, covered for the most part with large spreading trees, which afford a succession of leaves in all seasons. In those places where trees are scarce, a variety of flowering shrubs abound, most of them entirely new to a European, and surpassing in beauty, fragrance and number, all I ever saw in an uncultivated state: among these, a tall shrub, bearing an elegant white flower, which smells like English May, is particularly delightful, and perfumes the air to a great distance.

The fourth is from Mrs Elizabeth Macarthur:

> The greater part of the country is like an English park, and the trees give it the appearance of a wilderness or shrubbery, commonly attached to the habitations of people of fortune, filled with a variety of native plants, placed in a wild, irregular manner.

The fifth is from Surgeon Bowes:

> To describe the beautiful and novel appearance of the different coves and islands as we sailed up is a task I shall not undertake, as I am conscious I cannot do justice to the subject. Suffice it to say that the finest terras's, lawns, and grottos, with distinct plantations of the tallest and most stately trees I ever saw in any nobleman's grounds in England, cannot excel in beauty those which nature now presented to our view. (see Gilbert, 1962)

Yet despite the apparent range of responses expressed here, they are easily sorted, and a much wider sample would fall under much the same headings. The two unfavourable responses are utilitarian ('not one acre of clear ground': Ross and Clark, although military men, look at it with a farmer's eye, as if they might have to make a living out of it); two of the three favourable respond to the scene as picturesque (Tench and Bowes speak as comfortably salaried civil servants, who will *not* have to make a living out of it). This division has persisted to the present day, with one group taking the picturesque view ('superb and unique tract of *E. camaldulensis* riverine woodland') while the utilitarian has a bleaker response ('a muddy paddock with a few scruffy old gum-trees'). It is a misfortune that these two responses are so sharply distinct in Australia. Many of the finest landscapes of Europe are man-made, productive and yet harmonious with the natural environment. The Australian landscape, by contrast, was not as a rule immediately hospitable to 'human' – read 'British' – needs, so that utility and natural beauty have become sharply defined alternative categories. The heart of our environmental design problem, in my view, is to fuse those categories.

Another element in the responses quoted above that recurs again and again is the reason given for liking the landscape by Mrs Macarthur and Surgeon Bowes, that it is 'like an English park'; the rather open woodlands have an imported value attached to them, in that they suggest a noble estate. In time, the Macarthurs established 'Camden Park', and with it, a long tradition of such nostalgic place names across Australia for farms that often bore no resemblance to an English park. But why was the park-like aspect valued, and why did the landscape look like a park? In Britain, a park was land exempted from productive agricultural use, and thus the preserve of privilege and wealth. Parks were managed for hunting, riding, strolling. The trees were well spaced because they were mature, sometimes overmature: selective forestry thinned them out occasionally, for ease of riding, and the deer and other browsing animals kept the seedlings and understorey down. It was an intensely anthropogenic landscape, but seen as a natural one, in contrast with cropland. The irony is that the park-like woodlands of Australia, also seen as natural, were equally anthropogenic and equally the product of a hunting culture.

The comments by Major Ross that 'Nature is reversed' and 'worn out' are also much repeated. This is the Antipodes, where all things are topsy-turvy. The soils are 'worn out', and some of the landforms are old, but the flora has been evolving continuously to the present; it is certainly not 'ancient' in any sense, and since it is predominantly angiosperm, with fewer gymnosperms than any other continent, it is at least in that sense a 'young' flora. Its mammals are marsupials rather than placental, and thus represent an ancient lineage, but they are not, of course, Cretaceous mammals, or even Pleistocene ones (the Pleistocene fauna was quite different): they are the outcome of quite recent evolutionary events, as continuing palaeontological research has made abundantly clear.

When Major Ross called it 'worn out', he meant that it was strange, harsh, rugged, unlike the green and gentle, temperate homeland. If Australia had been settled by Spaniards, for example, the responses and behaviour might have been very different. Spanish (and behind them, Moorish) agricultural experience, respect for water, styles of architecture and of urban design might have led to a society more in harmony with the Australian environment, although it would probably have lacked the Anglo-Saxon virtues – for example, the talent for political stability – which also characterise our society. But, whatever might have been, it was the Anglo-Saxons (and the Irish) who came, thus generating the central paradox, that of a people whose cultural traditions and aspirations derive from a fundamentally different physical environment.

Of course the strangeness can be exaggerated; some visitors to Sydney in the 1820s remarked on the familiarity of the vegetation in its natural setting, because they knew the flowers already from English gardens (as I mentioned in Chapter 2). Visiting botanists generally had some knowledge of the flora before they set foot in Australia, and flowers from all over the continent were illustrated in works such as Robert Sweet's *Flora Australasica: or a Selection of Handsome and Curious Plants, Natives of New Holland, and the South Sea Islands – most Proper for the Conservatory or Greenhouse* (1827–28), drawn from flowering specimens in English gardens before Perth, Adelaide or Melbourne were settled (Gilbert, 1962). Thus the strangeness obviously depended on the experience of the observer.

The strength of the Anglo-Saxon cultural tradition can also be exaggerated. In some ways, behaviour was modified rather quickly to suit local conditions, for example, in the evolution of dry-farming techniques, yet the Anglo-Saxon conditioning was, and is, strong. The shortness of our pre-industrial history is also significant. By the end of the nineteenth century, Australians were beginning to accommodate themselves emotionally to their new setting; and physically, as illustrated by country towns like Beechworth, Clare, York and Bathurst. But evolving local tradition did not have long enough to establish itself firmly before it was overwhelmed by the homogenising technology of the twentieth century, in which so many forces are at work to override regional differentiation. Americans had stronger resources; for example, the landscape tradition established by Olmsted persisted into the twentieth century, and has been a base from which schools of landscape design have been able to tackle the new problems of a new century. Guilfoyle's work in Australia in the nineteenth century was as good as Olmsted's, but the skills died with him, and we have almost had to start all over again in the last decades.

Some of the dissonance between expressed perception of the environment in Australia and environmental behaviour is due to the speed of urban and industrial growth, so that the way we see ourselves and our setting is often at odds with the way we behave. The bronzed Anzac, the lifesaver, the great Australian athlete, the skilled bushman, lurks within us all, although by generally accepted health standards, Australians are among the least healthy groups living in 'developed' societies (Hetzel, 1974). We are aware of these inconsistencies, and many of our perceptions of the environment have undergone rapid evolution in the last few years: for example, the 1972 election, which brought Labor to power, marked the effective political recognition that most Australians lived in cities, and this recognition, if belated, will never again be neglected politically. Our perception of our cities themselves has also undergone rapid changes, as Bernard Smith shows – within a period of ten years, Paddington, for example, passed from an official slum, 'totally substandard', suitable only for demolition, to some of the most expensive real estate in Sydney (see Chapter 14). Our perceptions of our social environment are also changing very quickly.

Our capacity to adapt and evolve new responses is real, but we are still in trouble, I think. There are basic incompatibilities in our responses which we are reluctant to face. For instance, most Australians are perfectly sincere in saying they 'love the bush', including its fauna. To live 'in the bush' without destroying native animals is very difficult. Ecologists have been saying that the Australian environment is fragile for so long that 'fragile' is now a part of the popular vocabulary, but it does not seem to control basic thinking about our future in any significant way (for instance, population policy is still not a major political issue).

No other Western, highly urbanised and industrial country is as ecologically vulnerable as Australia. Advanced technology makes massive demands on natural systems. Flushing of pollutants, for example, depends on natural rainfall, yet the sum of the run-off from all Australian streams at their outlets is less than that of any one of the world's major rivers, for example the Mississippi. Our large cities are all in exceptionally favoured corners of the continent, but even in those corners flood, fire, and drought are recurrent realities, although we seem to be taken by surprise with every renewal. Both city and country life in Australia are dependent on massive power consumption, much of it currently drawn from non-renewable sources, as is the fertiliser on which Australian agriculture depends so heavily.

Much of our wealth, the wealth that has directly and indirectly supported our massive urban growth, has come from mineral resources, from gold in Victoria, silver-lead at Broken Hill, gold again at Kalgoorlie, through to the present boom. Geoffrey Blainey called his book on mineral exploitation in Australia *The Rush That Never Ended* (1961) but there could be a postscript, and I can imagine a complacent and well-fed New Zealand economic historian two hundred years hence writing an epitaph for Australia, as for Nauru: 'She was cursed with great riches in non-renewable mineral wealth'.

Although the riches are great, they are not inexhaustible, a fact to which our ghost towns bear witness. This is no argument against using them, but it would be unwise to base estimates of the permanent carrying capacity of the continent on current rates of resource consumption. Mineral wealth seems to engender a recklessly expansive view of the future, while the evidence that habitable Australia is *not* a big country is disregarded, even though it dominates our lives. We are, in fact, already short of land, and absurd though this may seem at first sight, we read about it in every newspaper (but with a different set of perceptions, not our 'Big Country' perceptions, but our 'Urban Problems' perceptions). When the newspapers report a shortage of land, we know that not enough building blocks are coming on to the market around the capital cities. We may then urge the government to make more land available. The outward expansion of the cities can be speeded up in various ways, but no government can make more land – except, perhaps, the Dutch. All, or almost all, the habitable land in Australia is already under competing pressures (for housing, for farming, for mining, for roads, for recreation), a fact starkly illustrated by the cost of land of any kind for any purpose within eighty kilometres of the major cities. An even clearer example of the shortage of land for specific purposes is the difficulty of finding an acceptable new site for almost any substantial enterprise in Australia; a new airport for Sydney, a container terminal, a new power station in Melbourne. Every new freeway site, major housing sub-division, sewerage works, dam site, quarry application is bitterly contested, usually with good reason, namely that the site already has valid uses, and the new uses would conflict with the old. The curious feature of this situation is that an acute community awareness of all these specific problems has not yet become a national perception that Australia is already short of land for some of its needs, which are primarily urban.

Because of its geology and latitude, Australia is short of water and good soils. These observations are commonplace. Compared with the other continents, Australia is poorly supplied with natural harbours, estuaries and wetlands. The limited supply of estuaries and wetlands presents us with acute problems in land-use, because they have a special biological function (as nurseries) that is incompatible with the technological demands we make of them. This conflict is to be found in all industrial countries, but it is exceptionally acute in Australia.

These are all essentially problems of urban and industrial land-use. Meanwhile, down on the farm things may be worse. Sir Keith Hancock has said of his own childhood:

> In *Country and Calling* I commented as follows on the privileges and deprivations of my boyhood in Gippsland: 'As a healthy young barbarian I found vivid joy with the Australian outdoors; but might not my joy have been deeper if somebody had taught me to read the story which time had written upon Australian earth?'. My teachers could not help me to use my eyes, because nobody had helped them to use their eyes. I wonder whether things are different today in Gippsland. (Hancock, 1974, p. 72)

A country boy may now be lucky enough to find people who know something of the land he lives in, but it is still likely to be an impoverished environment. One of the most telling comments for me in Donald Horne's *The Education of Young Donald* (1967) is his account of his upbringing in Muswellbrook, in the heart of the country, but cut off from nature. Horne learned about nature in the city, on his trips to Sydney, where he could walk the beaches and explore the Royal National Park and Port Hacking, much richer than the paddocks of the upper Hunter.

In *The Great Extermination*, a book published in 1966, a few years before its time and therefore without the impact it should have had, there is a chapter with the title, 'The decline of the plants', by John Turner. Writing of the basaltic grass-lands of western Victoria, he remarks that 'the native flora was virtually extermi-nated over hundreds of square miles'. He quotes an estimate of 2200 native plant species for Victoria, with 550 alien – 'about five new weed species established themselves every year between 1870–1930. This is an index of the progressive deterioration of the natural plant communities, for these are rarely invaded when in the natural state'. In fact, of the native plants, some 277 species are either extinct or survive only in isolated pockets. But the loss of individual species is less serious than the loss of whole communities:

69

vegetation was a mosaic of many different plant communities, dominated by different species of *Eucalyptus* throughout the better watered pastoral belt of Australia (White Box, White Ironbark, Yellow Box, etc.). Each had its characteristic topography, soil, climate, flora and fauna. With pasture improvement all these diverse communities are losing their individuality and are converging to one more or less uniform pasture, carrying a few exotic species of grass and clover – the surviving trees themselves are rarely allowed to regenerate; they are a diminishing asset. (Marshall, 1966)

Some Australians who 'love the bush' have in mind such cleared pastures with a few stands of remnant eucalypts. Many have no idea of what an undisturbed environment is like. I have heard a tattered few acres that has been logged, cleared and burned three times in a hundred years, and is now tertiary regrowth eucalyptus woodland choked with blackberry and watsonia, described and ardently defended as 'natural bushland'. I have been shown degenerate rainforest (a so-called national park) from which all of the valuable timber was removed over a century ago, beginning with the cedar-cutters, so that little remains today other than lawyer vines and the giant stinging nettle tree (*Laportea*), a weed species. This sorry relic was presented as pristine rainforest.

Many Australians know no better, because they have had little opportunity to learn. To be sure, they would have had no opportunity at all if we had never settled this continent, thus preserving the pristine environment in its entirety. Given settlement, of course the land had to be cleared, the trees felled, the native grasses ousted. Yet in Europe, the fields have been tilled for centuries without obliterating the indigenous flora. In England, and China or Japan, and equally in much of North America, the native flowers and grasses and shrubs and trees survive in meadow and hedgerows, and along the roadsides, as well as in woodlands and natural reserves. The difference lies in the isolation of the Australian flora, which rarely survives competition with introduced species in disturbed areas – that is, in most of agricultural and pastoral Australia – whereas the native flora of the temperate lands of the northern hemisphere can often co-exist with the species of agriculture, themselves plants of the northern hemisphere, with relatives among the natural communities of the regions in which they are cultivated. In Australia, the original communities are wiped out directly by clearing and indirectly by competition along the roads and elsewhere; where the original communities were rich and diverse, the flora that replaces them is very monotonous, consisting of a few agricultural species, together with their attendant introduced weeds, and it is in this sense that the environment is impoverished. Of course, much depends on the region. The young Keith Hancock had riches on his doorstep in east Gippsland, and his deprivation was the lack of a natural history tradition in his upbringing. But Donald Horne had neither the riches nor the natural history at Muswellbrook; and my own experience was closer to Horne's than Hancock's.

I spent a part of my youth in the Wimmera. I do not believe that I had a deprived childhood. It was a happy one for the most part, but it was deprived in three significant and interrelated ways. First, it was socially deprived. Like most country towns in Australia, ours was quite rigidly stratified socially. As I and a very few others went away to boarding school, I was not on holidays at the same time as the children who went to the local state school, and I hardly even knew them. This spelt alienation of one kind. I also had Hancock's problem of environmental education. Although I must have had a latent interest in natural history when I was young, there was no one to foster it, no one to tell me about the birds or the plants or the geology, or even the history of settlement of my own countryside. (We were familiar with names like Muller and Dahlenberg in the district, but knew nothing of the cultural history of the German settlers in the Wimmera, nor of their traditions of land-use, other than that they were respected as 'good farmers'.)

Finally, much of the Wimmera was and is an impoverished countryside. I do not recollect ever seeing a marsupial (except the possum) in its natural setting while I was young, and the natural vegetation had nearly all gone, replaced by exotics that serve our economic purposes better, but are very few in number of species, so that a complex ecosystem has been replaced by a monotonous and simple one. There was no Walden Pond at Horsham or Nhill, and could have been no Thoreau or Gilbert White, nor even Huck Finn. Most English and American country children enjoy much greater natural history resources than most Australian country children, except perhaps in part of the American mid-West, although there at least the corn is indigenous. If I seem to labour this point, it is because it is contrary to the popular view. Many English children have seen, sometimes even studied, rabbits and foxes, weasels, stoats, squirrels, water-rats, badgers perhaps, occasionally even otters. At least fifteen native mammals are to be found today within the confines of metropolitan London, and some of them, including the fox, mole, vole, shrew and hedgehog, together with the non-native grey squirrel, are common (Gill and Bonnet, 1973). Many English children know at least the popular names of the wayside flowers, and many of the birds, including those of the wetlands and the migratory species. This is partly the tradition of a culture biased to country pursuits, partly opportunity.

I grew up among paddocks of wheat, planted sugar gums from South Australia, weeds, a few surviving eucalypts, and the limited range of exotics common to country-town parks and gardens. The only animals with which I was familiar were dogs, cats, rabbits, horses, sheep and cows. There were birds, but I knew little of them, other than the magpie, and around the houses, sparrows, starlings, and mynah. None of this is to deny that the resources *potentially* available to me were richer than those of my English counterparts. I *could* have explored near-pristine bushland in the Grampians, and we did sometimes 'go for a drive' or even a picnic in country such as this, but it was not my day-to-day environment. I was obscurely drawn at this time of my life to a swamp down behind the town (Nhill, in this case) and to the Little Desert, long before it became a *cause célèbre*, but I did not know why, and never learned much about it. Almost the only other places where indigenous grasses and herbs were to be found in the area were the graveyards, but I did not know that either. I did not begin to learn until years later, and then at first out of books. I had to learn to begin to see Australia. Alec Hope's account of us as 'Second-hand Europeans, pullulating on alien shores' has been quoted too often, but it sums up much of our history to date. I think my experience is typical, and it is not therefore surprising that we have among the lowest standards of environmental design in the world. Australians are still learning to see where it is they live. The imaginative apprehension of a continent is as much a pioneering enterprise as is breaking the clod.

71

Postscript

Well, either things have got better or I more tolerant as I approach my appointment in Samara. It is hard, twenty years later, to grasp the excitement generated by that conference, among a group of very able people – but the issues remain alive as ever. We now talk about them more, but . . .

References

Blainey, G., 1961, *The Rush That Never Ended: A History of Australian Mining*, Melbourne University Press.

Blainey, G., 1966, *The Tyranny of Distance: How Distance Shaped Australia's History*, Melbourne: Sun Books.

Gilbert, Lionel, 1962, 'Botanical Investigation of Eastern Seaboard Australia 1788–1810', BA Honours thesis, University of New England.

Gill, Don, and Bonnett, Penelope, 1973, *Nature in the Urban Landscape: A Study of City Ecosystems*, Baltimore: York Press.

Hancock, W. K., 1972, *Discovering Monaro: A Study of Man's Impact on His Environment*, Cambridge University Press.

Hancock, W. K., 1974, 'Book Review of G. Seddon, *Sense of Place, a Response to an Environment, Canberra Historical Journal*, February, p. 72.

Heathcote, R. L. 1969, 'Drought in Australia: A Problem of Perception', *Geographical Review*, no. 59, pp. 175–94.

Hetzel, Basil, 1974, *Health and Australian Society*, Harmondsworth: Penguin.

Horne, D., 1967, *The Education of Young Donald*, Sydney: Angus & Robertson.

Hope A. D., 1973, *Selected Poems*, Sydney: Angus & Robertson.

Marshall, A. J. (ed.), 1966, *The Great Extermination: A Guide to Anglo-Australian Cupidity, Wickedness and Waste*, London: Heinemann.

Smith, B., 1960, *European Vision and the South Pacific 1768–1850: A Study in the History of Art and Ideas*, Oxford: Clarendon Press.

Sweet, Robert, 1827–28, *Flora Australasica: or a Selection of Handsome and Curious Plants, Natives of New Holland, and the South Sea Islands – Most Proper for the Conservatory or Greenhouse*, London: James Ridgeway.

Walmsley, D. J., and Day, R. A., 1972, *Perception and Man-Environment Interaction: A Bibliography and Guide to the Literature*, Occasional Papers in Geography, no. 2, Armidale: Geographical Society of New South Wales, New England Branch.

7 Eurocentrism and Australian science*

Despite its diverse origins in China, the Middle East and the eastern Mediterranean, the main development of science for the last four hundred years has been in western Europe, and this science has been coloured by European experience and social and cultural values. European bias is apparent in many ways that are so familiar that we take them for granted. That one of the two subcontinental appendages of the Asian land mass should still be referred to as 'the Continent' by Australian tourists is a trivial example of transferred English values. That we are 'the Antipodeans' is another: literally, of course, Europeans are equally antipodean to us (roughly, at least: the exact antipodes of Canberra is in the Atlantic Ocean, south-west of the Azores). The conventional orientation of the globe, with the North Pole on top and the South Pole below, equally reflects this view. I usually reverse globes when I pass them.

Early naturalists showed this conventional orientation when they regarded the flora and fauna of this continent as aberrant – a theme hinted at by Michael Hoare (1969) under the title '"All Things Are Queer and Opposite": Scientific Societies in Tasmania in the 1840s'. The naming of the platypus as *Ornithorhynchus paradoxus* is a clear example. There is nothing inherently paradoxical about the platypus, which is a highly evolved and well-adapted animal. In evolutionary terms, moreover, it is not only *not* paradoxical, but logical, almost predictable. Equally, there is no inherent reason whatsoever why swans should not be black rather than white, although the first black swans created a stir in Europe, not only in scientific circles, but equally among the philosophers. An even more comic example of unconscious Eurocentric bias is the use of the word 'discover' in all the 'new' lands. Major Mitchell's 'discovery' of Australia Felix must have been news indeed to the people whose ancestors had been living there for 30,000 years and more, just as Living-stone's discovery of the headwaters of the White Nile must have been illuminating for the people of the ancient kingdoms of Buganda and Bunyoro.

*First published as 'Eurocentrism and Australian Science: Some Examples', *Search*, vol. 12, no. 12, pp. 446–50.

These preliminary examples are relatively trivial, but there is a number of ways in which Eurocentrism has systematically distorted the progress of science in Australia, and sometimes still does so. I shall discuss four.

Colonial and metropolitan science

That our scientific culture has European (primarily English) roots is inevitable. It shows directly in many institutional arrangements – for example, the offshoot 'Royal Societies' in each state; the use of Kew Gardens as the major repository for much Australian plant material of taxonomic importance; and today by Australian universities' continued reliance for the assessment of their staff on publication in 'reputable international journals', which in fact means, as a rule, English or American journals (see Powell, 1980, for a detailed analysis).

But the assumption that Europe is metropolitan has shown up in more curious ways. Fifty years ago, textbooks on palaeontology showed the Late Paleozoic and Mesozoic marine organisms, related to the squid and *Nautilus* of today and known collectively as ammonoids, undergoing rapid speciation somewhere near Berlin and then setting out in wave after wave to the lesser corners of the globe. Until the early 1970s, the flora of Australia was described as having three elements; the auto-chthonous, the Indomalesian and the Antarctic. This concept begins with the great British botanist, J. D. Hooker, who visited Australia in 1840 as Surgeon-Naturalist on HMS *Terror*, in the course of a four-year voyage with HMS *Erebus* in the southern oceans. During the voyage, Hooker collected plants in South Africa, Tasmania, New Zealand, Tierra del Fuego and many remote islands. In his great six-volume work, *The Botany of the Antarctic Voyage*, there are two volumes on the flora of Tasmania (*Flora Tasmaniae*), which begin with a famous introductory essay on the plant geography of the southern temperate and Antarctic regions. He shows that while the 'Antarctic flora', as he called the southern beeches and their associates, are very uniform, there are other plant groups in each of Australia, Africa, New Zealand and South America that are substantially different. Hooker therefore hypothesised that during a warmer period there was a centre of distribution in the Antarctic from which the present southern beech flora radiated. This concept of course long preceded the theories of continental drift and plate tectonics.

When the subtropical and tropical rainforests of north-eastern Australia were studied, they too were seen to have close relatives elsewhere, in this case to the north, in Papua New Guinea, Indonesia and Malaysia. They thus became known as the Indomalesian elements, and were also seen as immigrants – in this case, as fairly recent ones. Only the sclerophylls, *Eucalyptus*, *Acacia*, *Grevillea*, *Hakea*, and *Banksia*, were seen as the 'Australian' or autochthonous element in our flora.

Although this account of the evolutionary origins of our present flora was backed up by a good deal of evidence, the case was far from complete; yet it seems to have had an acceptance that goes well beyond the evidence, and I suspect that this was an expression of unconscious colonial attitudes – the untidy bush is the real Australia, whereas the shady beech forests of Tasmania and parts of Victoria, and the subtropical and tropical rainforests of the east and north-east coast, must have been migrants from more exotic lands. It is now known that the 'Antarctic' and 'Indomalesian' elements were widespread throughout the continent during much of the Tertiary, and that the sclerophyll flora of today is of fairly recent origin, having evolved from proteaceous and myrtaceous rainforest ancestors. The reason that the southern beeches are distributed through South America, New Zealand and Tasmania is that these lands were all part of a single land mass. As for the 'Indomalesian' element, there is a great richness in 'primitive' angiosperms in north-east Queensland that has led some botanists to speculate that the angiosperms may have evolved here (see, for example, Webb and Tracey, 1981).

The above example illustrates the reciprocal of Eurocentrism, *the colonial stance*. It has been evinced in many other ways – our flora and fauna have often been described as 'primitive' whereas they are of fairly recent origin in their present form and at least as highly evolved as the flora and fauna of Europe, more diverse and generally more complex. If we turn to anthropology, this continent is now regarded as having been inhabited by man from 30,000–50,000 years ago or more, and thus has a very long record of human occupation. Before the Leakey discoveries in Africa, it was often assumed that most hominid evolution took place in Europe and around the Mediterranean basin. Some of it undoubtedly did, but not to the degree once taken for granted.

It might perhaps be said that Gordon Childe, the first internationally re-nowned archaeologist Australia produced, should bear some responsibility for hampering the development of this discipline in Australia. He left these shores in 1922, only returning to die in 1957, because he believed that the proper place in which to study archaeology was Europe. There he espoused the concept of a rather steady development of humankind through the ages, coining the terms 'Neolithic Revolution' and 'Urban Revolution' to explain the apparently sudden developments of agriculture and city life. These two great events, one following upon the heels of the other, were seen to pave the way for the Industrial Revolution of the eighteenth century (Cunliffe, 1973). His socio-economic ideas of man's development, overlain by his Marxist persuasions (or as he referred to it, Marxist perversion), are best summed up in the title of his little book first published in 1936, *Man Makes Himself*. However, while his theoretical framework for reviewing the history of humankind fitted well enough the pattern evinced by Europe (although it has since largely fallen from favour), it was thoroughly misleading when applied to the Australian Aborigines. This did not become apparent until fruitless years of work had passed in which Australian archaeologists and anthropologists tried to force the Aborigine into a European model of development. Indeed, the great break-through in the study of the history of the Australian Aborigines has come only since the theoretical framework developed by Childe was discarded in the 1960s by a generation of Australian archaeologists and anthropologists whose researches are now among the glories of Australian science.

The role of analogy

Much science begins from analogy, and Europeans began, inevitably, by judging from their own experience. Some analogies were profoundly misleading, such as the early presumption of a great inland river crossing the continent (Europe had three and the other continents all had one or more). Sir Joseph Banks was one of the first to make the assumption explicit: 'It is impossible to conceive that such a body of land as large as all Europe, does not produce vast rivers capable of being navigated into the heart of the interior' (Letter to the Colonial Secretary, 15 May 1778). Explorers who found

75

rivers like the Macquarie, which flows north-west, thought they had found the headwaters; the mouth was to be in the Kimberleys in Western Australia some 2700 kilometres away. Sturt in 1833 argued from stream projections, topography and climate that this was out of the question, yet as late as 1845 Major Mitchell was still looking for the Australian Amazon; as we saw in Chapter 4, he named Barcoo Creek in Queensland – actually the headwaters of Coopers Creek, which flows into Lake Eyre – the Victoria River, believing it to be the headwaters of the river of that name which entered the sea in the East Kimberleys, and the matter was not settled until A. C. Gregory's explorations in 1855–56.

Another view derived from European experience was that the bigger the trees, the better the soil, which led to heartbreaking attempts at settlement in the karri country in Western Australia, and the Otways and Strzeleckis in Victoria. Yet a further analogy was to lead colonists astray in searching for the cause of the 'York Road poison'. When stock was poisoned in the Swan River Colony, the plants that might have caused it were tested systematically, but the last to be tested were from the pea family, to which the culprits (*Gastrolobium* and *Oxylobium*) belonged. The legumes in Europe have few poisonous members, and the clovers, lupins and medics are among the most desirable pasture components. 'Leguminous plants are particularly suited for the food of animals and the human race', said Ludwig Preiss, a German botanist who was visiting Western Australia at the time. To prove his point, Preiss 'drank a wineglass of diluted fluid extracted from the leaves of York Road poison', *Gompholobium calycinum* (*Inquirer*, 17 March, 1841). Suffering no ill-effects, he recommended the plant to stockholders as 'the very best thing they could cultivate as artificial food for stock' (Cameron, 1977). Not only was the pea family well thought of; many species of *Oxylobium* and *Gompholobium* were palatable to stock, and they lacked the acrid taste and fetid smell of most English poison plants, which are either cyanogenetic or alkaloidal. The toxic agent in *Oxylobium* and *Gompholobium* is monofluoracetic acid: its presence in nature was not demonstrated effectively until 1943, but it is very potent, and went on to be used as the rabbit poison '1080'.

Geology illustrates the role of analogy from European experience very well. Werner, the German geologist, assumed that rock types such as the granite, shales, limestones and basalts of his native Saxony would be found around the world, and in this he was proved correct in general. He also assumed that they would occur in the same stratigraphic order (logically, given that he thought them all precipitates from a universal ocean). Some early geologists tried to fit the very different rock sequences here to the Wernerian sequence; and two episodes in the second half of the last century were especially perplexing – the dating of the coal measures of New South Wales, which are now known to be of Permian rather than Carboniferous age; and the dating of glacial sediments near Bacchus Marsh in Victoria and at Hallett Cove in South Australia.

Glacial sediments were assumed to date from the Great Ice Age, the Pleistocene Period, and the Australian examples were very puzzling, especially the South Australian ones; they were very fresh in appearance, yet at a latitude of only 35° south, remote from the mountain massif around Kosciusko, and at elevations that ranged from a little under 300 metres in the Mt Lofty Ranges almost to sea-level at Hallett Cove. Selwyn had discovered a glacial pavement in the Inman Valley in the Mt Lofty Ranges in 1859, an expanse of smooth rock that had been striated like the glacial pavements of Switzerland laid bare by retreating glaciers. Agassiz, the Swiss geologist, had published his influential work *Glacial System* only in 1846, and had spent the 1840s and 1850s in winning converts in Europe and North America to his theory of a Great Ice Age. It is therefore hardly surprising that the first glacials in Australia were automatically assumed to be of a similar age, especially when they were so fresh. Tate, in 1878, discovered glacial erratics, boulders and scree at Hallett Cove almost at sea-level, that looked almost as if the glacier had retreated quite recently. Some ingenious attempts were made to explain them. Two geologists, Scoular and von Lendenfeld, suggested that they must have been rafted north and dumped by an iceberg from much further south. Tate argued that either the climate must have been very much colder in South Australia than it is now, or that these sites must have been much higher, around 3000 metres at the time of the Pleistocene. He favoured the latter, but there was no independent evidence for this convenient assumption. It took nearly twenty years before the true age of these glacial sediments was established. They are Permian, not Pleistocene – part of a very widespread Permian glaciation which affected two-thirds of the continent. Glacial features of Precambrian age have also been identified in Australia. The South Australian glacials are so fresh because they were covered shortly after deposition, and not exhumed by erosion again until relatively recently (Scott, 1977).

The story of the coal measures is told in detail by Vallance (1975), who gives many other examples of a similar kind. The synopsis of his paper is as follows:

> One hundred years ago, when the Linnean Society of New South Wales began, geology in this country was a 'colonial' science – its base of authority still lay largely in Europe. For almost a century, geology in Australia has been dominated by concepts originating in Europe and transported, more or less uncritically, to a land being explored. European precedent is seen as having exerted, in many cases, a counter-productive influence on geological progress here in the years before 1875.

Australian equivalents of the Old Red Sandstone, the Mountain Limestone, the New Red Sandstone and other classic European formations were identified by a variety of geologists, and the coal at Newcastle was of course assigned to the Carboniferous. The major problem was that geology was itself in flux, and all of the European controversies were reflected in this country, but they persisted in misdirecting attention here longer than they might have done. The 'classic' European formations represent a particular set of sedimentary, volcanic, plutonic or metamorphic events at a particular place at particular times, and there was no reason whatsoever why similar circumstances should have prevailed in a different corner of the globe at the same times. Even when the habit of finding equivalents of European formations was finally put to rest, the stratigraphic divisions and the geological time-scale continued to create problems and difficulties for geologists working out of Europe.

For one thing, all the classic 'sections' were in Europe – the Cambrian, Ordovician and Silurian were in Wales, and the Permian in Russia was the furthest afield. Although some fossil organisms were cosmopolitan, many were not, and our Permian coal flora was quite unlike the plants of Permian Europe, although closely related to those of India and Africa. Moreover, the major divisions were all separated by unconformities, which indicate breaks in the continuity of sedimentation, and hence *gaps* in the record. Such breaks were not likely to occur at the same times elsewhere, and bits of the record kept on turning up with no equivalent in the classic sections, because they represented the unconformities. This example goes beyond the role of analogy; it illustrates the way in which a classificatory system based on European experience was ill-suited for world-wide use. It has taken more than a hundred years of laborious emendation to reduce its deficiencies.

Explanation of the unfamiliar by the familiar

Only the unfamiliar calls for explanation. Thus scientists in Australia have spent nearly a hundred years in considering the adaptive functions of the sclerophyll leaf, exemplified by the eucalypts. In fact, such semi-rigid, leathery leaves are common from rainforest to savannah. The leaf type that is relatively rare is that of cool temperate trees like the maple: deciduous, bright green, rather soft and floppy, stiffened by little sausage-like cells pumped full of water and thus capable of wilting when water is scarce (Seddon, 1974). The science of ecology probably owes its origins to the great simplifications wrought in the flora of Europe by the Ice Age – the resulting ecosystems are reduced and therefore simple enough to study, but they are not at all typical of the world at large. Neither are the many lakes; nor the soils, rich in nutrients, rejuvenated by glaciation.

We might remind ourselves in passing that industrial technology therefore grew up in an exceptionally endowed, resilient environment, but it has been transferred to lands that are inherently more fragile, of which Australia is one.

Classification

These are ways in which the practice of science in Australia has been shaped not by universal truths but by the bias inherent in European experience and preconceptions. Our classificatory systems are all European in origin. They are in part based on analogy, and my discussion of the stratigraphic system arose out of a discussion of false analogies in geology, but they go beyond analogy. One of the most interesting and significant examples of Eurocentrism in science – one that can still actively distort the perception of reality in its field – is to be found in the taxonomy of the flowering plants. This might have been very different, both in outline and in detail, had it developed from another culture, or another place, such as Australia or New Zealand, and yet still have served the needs of botanical science as efficiently as the system we now have.

That system is Eurocentric in two quite distinct ways. First, the whole idea of genus and species is a natural product of European thought of the seventeenth and eighteenth centuries, based on the current understanding of Aristotelian logic (Cain, 1958). Thus the logical basis of the classification is derived from a particular place at a particular time, and its hierarchical character, from species at the base of the pyramid, through genus and family to kingdom, is a reflection of basic assumptions about social and natural order. This hierarchy later was to be interpreted rather as a reflection of the pyramid of evolution, but that was not its origin, and the formal hierarchy has inhibited research into evolutionary phenomena that do not fit this pattern – for example, convergent evolution, parallel evolution, interactive evolution, and polymorphic origins for new groups, all of which have a non-pyramidal geometry.

78

Landprinting at the coastline: feeding patterns of small crustaceans that ingest the sand in the swash zone, extract the organic detritus and evacuate the mineral content as pellets. JOHN HANRAHAN

Men . . . are able to suckle infants: 'Birth of the
Australian', 1996, oil on canvas, 153 x 102 cm.
JULIA CICCARONE

There are few other places in the world where the colours of the sea are so startlingly lovely: dolerite coastline at West Cape Howe (West Cape Howe National Park), 30 km south-west of Albany, WA. JOHN HANRAHAN

The Snowy at New Guinea Bend in Victoria some 60 km
from the sea: cliffs of the Buchan Limestone are in the
background, with rainforest at the base. 'New Guinea' is
a foresters' name, prompted by the dense vegetation.
DAVID CALLOW

Below the Snowy (or Jimenbuen) Falls in one of the
least accessible gorges of the river about 25 km north of
the Victorian–New South Wales border.

Aerial view of a few blocks of Melbourne, oriented to
the points of the compass: middle-class landprinting in
south-east Australia. GEORGE SEDDON

The main street in Yackandandah in north-eastern
Victoria. Most Australian towns once had verandahs
and street trees, later sacrificed for the car. There is
nostalgia in admiring Yackandandah ('Yack' to the
locals), but the trees and awnings are defensible on
functional grounds, and the form serves to unify the
streetscape and to give a comfortable sense of continuity with the past.

Sandhills near Albany, WA, patterned by wind and water along a high energy coastline.
JOHN HANRAHAN

A ghost gum *(Eucalyptus papuana)* in the MacDonnell Ranges near Alice Springs in Central Australia. This species is a bloodwood (subgenus *Corymbia*) like the marri of the gumnut babies. It is perhaps the most often painted and photographed of all Australian eucalypts, partly because its slim white trunk is so often limned against blood-red cliffs. JOHN HANRAHAN

The gumnut babies, rear view. MAY GIBBS

A pool in a narrow gorge in the Bungle Bungle
National Park in north-western Australia, vibrant with
colour. GEORGE SEDDON

But further, the Linnaean system was applied to a very restricted range of plant material, and this is the basis for the detailed classification. A number of very interesting consequences follows from this restriction – some two-thirds of all the genera in Linnaeus' *Species Plantarum* (1753) are European, and this work, we should remember, gives an account of all species of the plant kingdom known to science a little over two centuries ago. Many of these consequences are worked out by Max Walters, the Director of the University of Cambridge Botanic Gardens, in a thoughtful paper (Walters, 1961) that has not had the attention it warrants. I shall give a brief account of only two of his many examples.

First, he remarks that any student of angiosperm classification will know that there are, broadly, two kinds of families, those with relatively sharp boundaries, like the Umbelliferae (carrot family) and those that are distinctive in their typical, central members, but grade off into others at the margins, like the Rosaceae (or Rose family). In his introduction to the Rosaceae, de Jussieu (*Genera Plantarum*, 1789) says that: 'Tournefort gave the name Rosaceae to all those plants with regular polypetalous flowers which were not Umbelliferae nor Cruciferae, nor resembled Lilies nor Dianthi in their flowers'. In other words, they are what is left over after the sharply definable related families have been cordoned off. Why then is the rose chosen? The answer, says Walters (1961, p. 78), is clear.

> This was happening in seventeenth century Europe, where for centuries previously art and literature had been full of certain symbolic flowers. How could any other choice have been made? The 'indefinable' families, then, are associative; the type genus is an important European plant; and the shape of the family is a product of this thought process. Furthermore, the more powerful the symbol in medieval writing, the earlier the 'recognition' and the larger the family; thus Rosaceae and Liliaceae in contrast with (say) Hamamelidaceae and Amaryllidaceae.

One has only to reflect that the great Rose family is represented in Australia only by a few insignificant plants such as the unwelcome Bidgee-widgee, *Acmena anserinifolia*, to realise that a classification derived primarily from Australian plant material would have been very different at the family level. The Lily family was even more oddly represented, by the bizarre grasstrees, although these are now assigned to a family of their own, the Xanthorhoeaceae, and they are bizarre only if you think of them as failed lilies.

Note that in his comments above, Walters claims that the 'older' plant families (that is, families that were early named and recognised) are also the larger ones (judged by the number of genera they contain), and he goes on to show why this was so by historical analysis. Conversely, all families with only one genus – such as the Cephalotaceae, which contains only the Western Australian pitcher plant – are of relatively recent origin. Such families could not 'arise' early because of the nature of the thought processes of early taxonomists, who were concerned with arranging their 'kinds' of plants in a number of groups possessing common characters and of convenient size. There was no claim that the classification was 'natural'; it was intended, rather, to be *useful*. Note that de Jussieu used exactly one hundred family names; this was a good round figure.

Thus Walters argues that claims by later botanists that the taxonomic stability of many of these families and genera is evidence that they have a 'natural' validity ignore the realities of long cultural reinforcement. One botanist, J. D. C. Willis (1949), has argued that the size of a family in numbers and distribution is proportional to its age in evolutionary terms, and that the very small and restricted families are generally of fairly recent origin. He would thus appear to mistake a classificatory artefact for an evolutionary pattern. This is in some ways the most interesting example of this kind of thinking in that its effects are still operative, although I believe that there are others.

A related theme has been voiced by many European botanists who have spent much of their working lives in the tropics, of whom Emeritus Professor E. J. H. Corner of the University of Cambridge is one of the best known. His basic theme is that the trees of the tropics, and not the annuals of Europe, are the typical plants – morphologically, anatomically and ecologically. For example, the shape of tropical trees is often determined not by their way of ramification (the branching pattern) but by their selective abscission (the way in which leaves, pinnules and branches fall off). Branch abscission is fairly common in many tropical trees, and is common among some eucalypts; the lemon-scented gum (*Eucalyptus citriodora*) is a typical example, but its habit of clean branch fall is not found among the oaks, elms and beeches of temperate Europe. Tropical leaves often also show abscission in parts along the rachis (defying European rules in school textbooks); the pattern of leaf-fall in the Queensland umbrella tree (*Schefflera actinophylla*) is an easily available example of abscissing leaflets. The 'temperate' fall of entire leaves, and leaves only, is secondary (Van der Pijl, 1957). This is a further example of Eurocentrism that is not limited to Australia, in that school botany texts generally are based substantially on familiar European examples that are not, however, typical of the plant world at large.

Conclusion

These examples of various kinds may raise questions such as the following: What else could or should the early scientists have done? Were they at fault? Is Eurocentrism a historic phase only, or is it still operative? Does it matter?

The answer to the first question varies with the case. The assignment of glacial sediments to the Pleistocene, for example, was inevitable. It is both usual and proper to interpret field evidence in the light of current theory. When it does not easily fit the interpretative framework, it is usual either to question the validity of the data (which many did in this case) or to look for plausible *ad hoc* explanations, like the drifting iceberg. It needs more than one set of anomalous data before we challenge our interpretative framework. The colonial aspect is reflected most in the time lag. A due application of the stratigraphic principles evolved by 'Strata Smith' in England and practised by eminent geologists such as Lyell were well established, and should have resolved this problem much sooner, as it would have done had these glacials been found in Europe. But then scientists of any kind were few in colonial Australia, and rarely of the first rank, so the delay is not really so strange.

Much of the same might be said of the stock poison in the Swan River Colony. It is sensible to test the most likely agents first, and leave the least likely to the end. Science is not so empirical that everything and anything is tested in random order. My account of the historic examples is therefore more descriptive than pejorative.

There are nevertheless some Eurocentric aspects of science in Australia that are to be deplored, and very often these are the ones that are still with us. The deference to institutional structures dominated by the northern hemisphere is one example, of which publication in international journals, mentioned above, is only a part. The theoretical framework within which scientists in Australia work is generally imported. Although it is again inevitable in great part, it has tended to turn many Australian scientists into sooty empirics or slaves to data accumulation, with little interest in scientific theory, although theory is the heart of science (the excessive emphasis on the 'information content' of science and on data accumulation is, I think, one of the reasons why science teaching in Australian schools and universities is often so dull).

There is, however, an interesting corollary to this: much of the more valuable work in science in Australia has been the outcome of the rejection of an imported framework that just could not be made to work. The glacial story illustrates this point, in that one response to Tate's discoveries at sea-level was to send other geologists to the areas where one might reasonably *expect* Pleistocene glaciation, in the Australian Alps and in Tasmania. It was then established that the Pleistocene glaciations were very restricted in Australia. When the Permian age of the Hallett Cove glacials was identified and linked to other Permian examples in Australia, it did not take long to relate them to similar glacial sediments in India. This became part of the slow accumulation of data that were finally to be explained by sea-floor spreading: India had adjoined the west coast of Australia at the time, and the two sets of glacial phenomena were part of a single event.

Research on brown coal in Australia has had a somewhat similar history; the theoretical approach was determined in Europe–North America from an experience based primarily on bituminous coal. For some years this was a significant distortion of coal research in Australia, in that the framework was not appropriate to the study or the technology of brown coal. This then led to some spectacular technological failures. For example, the original Yallourn boilers were designed for coal of 50 per cent moisture content, whereas Yallourn coal has 67 per cent moisture. The reason for the mistake was that German coals being worked at the time had 50 per cent moisture, and so did the brown coals from Yallourn North, but they were atypical of Yallourn coal. The consequence of the mistake was that the boilers didn't work. When the Yallourn C plant later turned to pulverised-coal boilers, there was still little knowledge of the burning characteristics of these moist coals, and unstable combustion led to pressure waves that at times came near to explosion. Eventually, after some thirty years, the differences were productive, in that the inherited framework was abandoned and a more appropriate one was worked out locally, and in the course of this work some very significant advances were made in the understanding of the physics and chemistry of coal (Evans and Allardice, 1978, p. 82).

In his paper on the basis of Linnaean classification discussed above, Cain puts the view that botanists would do well to reconsider the theoretical bases of Linnaean taxonomy:

> They should be examined for two reasons. In the first place there is no better method for scientists of one period to bring to light their own unconscious, or at least undiscussed, presuppositions (which may insidiously undermine all their work) than to study their own subject in a different period. And secondly, when the writings of an earlier author have apparently been taken as the basis of subsequent work, constant scrutiny is necessary to prevent his presuppositions becoming fossilised, so to speak, in the subject and producing unnoticed inconsistencies when modifications have been made as a result of subsequent work. (Cain 1958, p. 44)

81

What Cain says of differences in time is applicable also to differences in place, and the difference in perspective that comes from inhabiting this still remote southern continent can have some advantages. It should be a little easier for Australian scientists to identify 'fossilised presuppositions' that have been built into the structure of science because of the accidents of its origins.

Postscript

This essay is the only sample of my 'History and Philosophy of Science' phase. I have explored these themes more fully in the papers set out in the following list of references.

References

Cain, A. J., 1958, 'Logic and Memory in Linnaeus' System of Taxonomy', *Proceedings of the Linnean Society of London*, no. 169, pp. 144–63.

Cameron, J. M. R., 1977, 'Poison Plants in Western Australia and Colonizer Problem Solving', *Journal of the Royal Society of Western Australia*, vol. 59, no. 3, pp. 71–7.

Childe, G., 1936, *Man Makes Himself*, London: Watts.

Cunliffe, B., 1973, 'Introduction', in V. Gordon Childe, *The Dawn of European Civilisation*, St Albans: Paladin.

Evans, G. D., and Allardice, D. J., 1978, 'Physical Property Measurements on Coals, Especially Brown Coals', in Clarence Karr Jr (ed.), *Analytical Methods for Coal and Coal Products*, New York: Academic Press.

Hoare, M., 1969, ' "All Things Are Queer and Opposite": Scientific Societies in Tasmania in the 1840s', *Isis*, no. 60, pp. 198–209.

Hooker, J. D., 1847–60, *Flora Tasmaniae*, vols V and VI in *The Botany of the Antarctic Voyage of H. M. Discovery Ships* Erebus *and* Terror *in the Years 1839–1843*, London: Reeve.

Powell, J. M., 1980, 'The Haunting of Saloman's House: Geography and the Limits of Science', *Australian Geographer*, no. 14, pp. 327–32.

Seddon, G., 1971, 'Logical Possibility', *Mind: A Quarterly Review of Psychology and Philosophy*, vol. 18, no. 324, October.

Seddon, G., 1973, 'Abraham Gottlob Werner: History and Folk-History', *Journal of the Geological Society of Australia*, vol. 20, part 4, pp. 381–95, December.

Seddon, G., 1974, 'Xerophytes, Xeromorphs and Sclerophylls: The History of Some Concepts in Ecology', *Biological Journal of the Linnean Society*, vol. 6, no. 1, pp. 65–87.

Seddon, G., 1996, 'Thinking Like a Geologist', *Journal of Earth Sciences*, vol. 43, no. 5, pp. 487–95.

Scott, H., 1977, *The Development of Landform Studies in Australia*, Artarmon: Bellbird Publishing Service.

Vallance, T. G., 1975, 'Origins of Australian Geology', *Proceedings of the Linnean Society of New South Wales*, no. 100, pp. 13–43.

Van Der Pijl, L., 1957, 'The Tropical Face of Botany', *Proceedings of the Eighth Pacific Science Congress of the Pacific Science Association*, vol. IV, pp. 342–3.

Walters, S. M., 1961, 'The Shaping of Angiosperm Taxonomy, *The New Phytologist*, no. 60, pp. 74–84.

Webb, L. J., and Tracey, G. J., 1981, 'Australian Rainforests: Patterns and Change', *Ecological Biogeography of Australia*, 2nd edn, The Hague: W. Junk.

Willis, J. D. C., 1949, 'Birth and Spread of Plants', reprinted from *Boissiera*, vol. 8, Waltham, Massachusetts: Chronica Botanica Co.

Figures in the landscape*

Many of the images discussed here consist of a view of the City of Perth from Mt Eliza, Kings Park, which has always been the historic vantage point for this city, and still is: the people of Perth periodically renew their sense of possession of their city by going up to the park to look out over it. The view has also always had a promotional role. Visitors are taken to see it, and many of the early paintings and photographs were made to reassure those at home of the wisdom of the emigrant's choice. But it does not matter if you are not familiar with Perth. The subject of my exploration is not the contents of the view, but the ways in which it has been projected.

In most landscape portraits painted, sketched or photographed in the nineteenth century in Australia, there is an implied relation between 'man and all his works', and their setting in a new land. If this is the subject of the portraits – the relation between our species and our environment – then the object is to show that this 'new' land has been humanised (earlier inhabitants not counting). Sometimes, we are represented only by our works, as in the 1836 painting by Thomas Turner of *Albion House*, Augusta, which shows a landscape dominated by eucalypts, with some rough – and obviously arduous – clearing in the foreground, then a well-tended field extending to the inlet down slope from a simple Georgian house, backed by more eucalypts. The painting is a proud record of achievement in a setting that bespeaks loneliness and courage (see Chapman, 1979).

Sometimes, we are represented by our domestic animals, cows grazing in the bush, as in some Heysen paintings. But we are usually there *in propria persona*, usually in the foreground. Both house and cow announce a common theme, that of *domestication*, in a setting that announces its antithesis. The theme is a variant of one that was common to the Romantic tradition of painting in Europe, exemplified, for example, in the work of the influential German painter, Caspar David Friedrich (1774–1840).

*First published as 'Figures in a Landscape', *Landscape Australia*, vol. 9, no. 2, 1987, pp. 107–14, and reprinted in G. Seddon and D. Ravine, *A City and Its Setting: Images of Perth*, 1987, Fremantle Arts Centre Press.

There are figures in nearly all the views from Mt Eliza until the 1930s, when they disappear. The disappearance is not total – there were always a few landscapes from earlier times without figures, and a few later ones with figures, but most landscapes up to the 1930s have figures, and most post-1930s landscapes do not. The figures serve a variety of functions, in relation to formal composition, and to content and mood. As compositional elements, the figures may be used to provide scale and balance, always off-centre, commonly on the right hand side. The figures are often in silhouette, with their backs to the painter or camera, and they constitute a foreground frieze that works like a stage set in giving a better sense of depth, and also of involvement, in that they are doing what we are doing – looking at the view.

Up to the 1850s, the foreground figures are often Aborigines, who have an emblematic function, as do black swans and kangaroo paws. We are certainly not being asked to see the invasive settlement through the eyes of the displaced black observers, and there is a striking opacity of vision in using figures in this way without asking what they saw. They are as blind as the stone Queen Victoria gazing imperiously from her pedestal over Perth Water to the young city.

James Walsh, an ex-convict, painted in the 1860s a remarkable series of Aboriginal studies that are neither romanticised nor caricatured, but rather a lively record of daily activities: making fire, spearing fish in the Swan River and so on. He also painted *View from Mt Eliza*, which has a group of five Aborigines on the left around some bark sleeping shelters, rendered with all the ethnographic detachment of his other native studies – but they are approached from the right by a woman in crinoline and pelisse, large hat, and walking stick held out before her. This is not an image of confrontation, however, in that the natives are paying no attention to the woman, nor she to them. Walsh appears to be portraying two worlds that barely recognise the existence of one another, but his intentions can only be guessed at. The effect is eery.

The similarity in formal composition highlights the contrast in mood and content between successive views from the same vantage point. A photograph from the early 1860s by A. H. Stone shows his daughter Lizzy in the right foreground, a delicately dressed little girl with tight-waisted jacket, full skirts and graceful straw hat. She is in reverse silhouette, spotlit by a low western sun against the sombre bush setting, and she looks fragile and isolated.

A 1905 photograph from Mt Eliza is very different in mood, but very similar in composition. Two dark figures, right foreground, with their backs to the camera, look north-east towards the city across Perth Water. The difference in mood is that the foreground is an English garden with daisies and other flowers in formal beds; and the figures are schoolboys in knickerbockers and jackets with midshipmen's collars; the Anglicisation of the environment is thus complete, and the image has become artefact, in that the vision of an English town in an English setting, implicit in many idealised views of the 1840s and 1850s, has now been substantially realised 'on the ground'.

A more elaborated use of human figures, but one that is still basically static and theatrical, is what might be called the 'tableau vivant', in which a scene is represented by posing appropriate figures against a striking backdrop; for example, a photograph in 1911 taken from Mt Eliza with the city in the background. The foreground is all artefact, with a kerbed gravel terrace around a statue of Queen

Perth from Mt Eliza, 1983.
MICHAL LEWI

View from Mt Eliza, anon. 1842. The Aboriginal figures, in silhouette against Perth Water, have an obviously compositional function: the left figure has his arm extended in a natural gesture to the little town below. He is clearly addressing his companion, but we are not encouraged to speculate on the content of his speech, unless it be that 'All this makes our extinction well justified'. The grasstrees in the foreground and other details show an attempt at accurate representation – limited by a lack of skill and of appropriate conventions for representing the eucalypts. The tree on the right reappears as a dead stump in photographs of the 1860s. MITCHELL LIBRARY

Queen Victoria's view from Kings Park in 1911: a 'tableau vivant' of established social order and the emblems of Imperial power in an environment dominated by the man-made. The figures are skilfully and naturally posed to make a group portrait, yet they do not dominate the scene, so they remain figures in a landscape rather than a portrait with a landscape background, although they are facing the camera rather than the view. BATTYE LIBRARY NO. 22P

Victoria, who is admiring the view (of one of the lesser jewels in her crown). The figures in the foreground, of schoolboys, attendant matrons and promenading gentry, are arranged in the receding planes and balanced composition characteristic of this genre.

An urbane tableau vivant is given in a coloured lithograph drawn from the Barracks in the 1870s, looking down St Georges Terrace, but with a 180-degree panoramic sweep right round to Mill Point and Mt Eliza. This scene is full of figures, intended as a record of activity – men walking down the road, cows in an enclosure, boats on Perth Water, and work and play of various kinds in progress. This is in a popular late eighteenth- and nineteenth-century tradition of the urban panorama. An 1801 view of Trieste at the head of the Adriatic has much in common with many of the views of Perth from Mt Eliza in the nineteenth century, especially the broad curving bay, the hills in the background, the ships at anchor and the paved esplanade between town and water, the Trieste equivalent of Mounts Bay Road.

We began with figures in the landscape who constituted little more than a foreground frieze, and moved on to the 'tableau vivant' in which the figures are arranged to make a narrative point. There is thus a continuum in which the emphasis moves steadily from the landscape to the figures. The schoolboys and Queen Victoria are intermediate, in that the emphasis is more or less equally distributed.

The portrait with a landscape background is, of course, a common genre, and many people have gone up to Mt Eliza to be photographed, the message of such a photograph being that 'we were there'. An exuberant example is that of three demure gentlewomen sitting on the base of a World War I cannon, with a more adventurous and younger woman sitting on top, showing an ankle. Perth Water is in the background. A more formal example is the photographic record of Sir John Forrest planting a tree on Mt Eliza in 1895, where the record of the event is the function of the photograph, and the Swan Estuary is merely a backdrop. The figures dominate the landscape in this example.

A rather different use of figures in the landscape is shown in several well-known views from Mt Eliza, including that by Horace Samson in 1847, and an engraving from the *Picturesque Atlas of Australia*, published in 1888. Neither of these show foreground figures, but they both include small figures in the middle distance walking along what was to become Mounts Bay Road; in the 1888 view, there are also several horse-drawn drays. These small figures in the middle distance are reminiscent of Chinese landscape painting, which is rarely without this device. The figures are more dynamic than foreground figures – they are almost invariably moving further into the picture, carrying the observer's eye with them; and, of course, they humanise the landscape.

There are a few early views of Perth without figures of any kind, but these seem to belong to a somewhat different tradition, that of objective record, a kind of pictorial surveying like the topographic delineations of coastline kept by Matthew Flinders and others engaged in coastal survey. There are also some views in which there are figures almost by accident – one has the feeling that they were neither sought nor avoided, but just happened to be there, like the view from Kings Park that includes a little dog and a child with a hoop running along Mount Street in the 1870s.

But, from around 1935, the figures go, by conscious avoidance. There is a particularly good series of postcards taken by professional photographers (the Rose series) that appears during these years, and there are no figures in them. This then becomes a standard format of bushland foreground, Perth Water and the city beyond it.

The odd exception to this format looks naive; for example a 'postcard' view as late as 1967 shows a lawn setting in the foreground, lemon-scented gums, a park bench, and nine very middle-class-looking people in the foreground, all but one of them female, and five of them well-turned-out young girls. Five of the nine are fair-haired. This photograph is probably not posed, and no one is looking at the camera, but the photographer has assembled the material to hand to make a clear statement. The bush has been utterly tamed; even the gum trees are tidy, and this is a world made safe for suburban middle-class values. The photograph has a very old-fashioned air, representative only of Western Australia or Queensland as late as 1967. The standard views of central Sydney and Melbourne at this date are of the Harbour Bridge and/or the Opera House for Sydney, and a view of central Melbourne framed in plane trees, across the Yarra from Alexandra Parade, and there are no people in either.

What has become of the figures? It is easy to observe both amateur and professional photographers at work in Kings Park. They wait until any passing pedestrians are out of the way before they shoot. Moreover, people who see that a photograph is about to be taken step back and keep out of the field of view. They know they don't belong. The only exception is when a 'we were there' photograph is being taken, usually by a happy and unselfconscious group of country people without much experience of travel.

The 'we were there' photograph can be illustrated very clearly by a visit to an international tourist target such as Borobodur, the great Buddhist temple in central Java. There, the Indonesian visitors merrily photograph one another with the temple as a blurred background, while the French and German tourists fiddle seriously with lenses, make exposure and depth of field calculations, and wait for a clear view: all temple, no people. What is happening here is obvious enough – specialisation of subject, either a portrait of a landscape, or of a person or group of people.

But that can hardly be the whole story. We are recording a shift in convention, which may have no meaning at the individual level, because individual painters and photographers first put the figures in and, later, left them out, simply because it was a convention. But why has the convention changed? Do we no longer wish to humanise the landscape? Do we, in fact, regard it as over-peopled, and therefore wish to see 'the thing in itself'? Have we accepted the Australian landscape, and no longer feel the need to humanise (or Anglicise) it?

In sketching the dimensions of an answer to this question, we could turn to the history of landscape painting. There are almost always figures in the landscape paintings of the European tradition in the eighteenth and early nineteenth centuries. Sometimes, they are consciously picturesque, as in the landscapes of Salvator Rosa and Claude Lorrain, of which the vignettes from the Tallis map of 1850 are a distant echo. Sometimes they are 'honest scenes of workaday country life', as exemplified in such different painters as Constable and Millet, and, in Australia, Tom Roberts. Sometimes, there is a strong narrative interest, as in many of the paintings of McCubbin, such as *Lost* and *Down on his Luck*. These paintings are a strong expression of concern for man in his setting, and create a mood somewhat akin to that of Thomas Turner's *Albion House* with which we began.

The first major English landscape painter to omit the figures was Turner, clearly because his subject is so often the interplay or conflict of natural forces – wind and water and, above all, light – and human figures play no part in this. Corot has had a major influence on landscape photography, and he rarely included figures. The French Impressionists usually had figures, especially Seurat and Renoir, whose sensuous landscapes are perfumed with and complemented by female flesh. Cézanne painted landscapes without figures, but perhaps they are not landscapes either; the nominal subject becomes translated into structure and colour.

In Australia, Streeton sometimes painted landscapes without figures, but his landscapes are nevertheless humane, and they seem to be adapted to civilised purposes. Henry van Raalte was painting landscapes without figures in Western Australia in the early 1900s; he appears to be a link from Corot to Heysen. Heysen painted gum trees, and his landscapes are usually without people: he was having a love affair with tree trunks.

Later Australian painters have often continued to include figures. Tucker, Boyd and Nolan do so to mythologise the landscape; Drysdale to offer a social commentary on people and the land and their influence on each other. Fred Williams painted unpeopled landscapes, continuing in his own way the topographic surveys of Flinders.

The great French photographer, Eugène Alger (1857–1927) recorded doorways, courtyards, the buildings, streets and alleys of Paris, all without people, and this has become general practice for townscape and architectural photography. The professional photographers of landscape – and Australia has some great ones, such as Colin Totterdell and Olegas Truchanas – record empty landscapes, preferably with no trace of the hand of man.

Some photographers of natural scenes say that they avoid figures because figures 'date' the photographs, and this may be relevant to the Rose series postcards, which were marketed nationally and had a fairly large print-run. But this change is not restricted to commercial postcards: it is equally characteristic of the work of amateur photographers, and therefore still requires an explanation beyond the commercial motive. Figures do, of course, 'date' both paintings and photographs. In doing so, they often add greatly to their charm and interest. Moreover, landscapes without figures can also be dated, both from style and content – it is possible to date any photograph from Mt Eliza within a couple of years. The interesting question therefore is to ask why photographers *want* their landscapes to be timeless.

Speculation is easy, and might yield a dozen possible explanations, none of them altogether convincing. What does seem clear is that our sense of ourselves in relation to our environment has changed in some way. We began by describing a continuum of figures and landscape, with the emphasis at first on the landscape, with the figures as little more than a decorative frieze, and at the other end of the continuum, the emphasis on the figures, with the landscape functioning only as a backdrop. We seem now to have gone off the scale at both ends, to specialise more firmly in subject matter: either we photograph people or a landscape, independently. We wish to see the landscape in its own right. People in the foreground are a distraction because they particularise too strongly – we see their dress, which indicates the fashions of the year, even the season; and their bearing, which places them socially and economically. Perhaps television and the cult of personality have wrought this change, and made people too individually specific in the detail of their appearance.

Postscript

This essay is inconclusive. It asks a question and finds no answer. I still don't know the answer, and I still find the question interesting.

References

Chapman, Barbara, 1979, *The Colonial Eye: A Topographical and Artistic Record of the Life and Landscape of Western Australia 1798–1914*, Perth: The Art Gallery of Western Australia.

9 Dreaming up a rainforest*

Northern Europe has had three persistent dreams of place: the first, of the Mediterranean; the second, of coral islands in the Pacific; and the third, of tropical jungle. All three are essentially romantic, and were especially potent in the nineteenth century. The Mediterranean dream is in Byron, Shelley, Norman Douglas' *South Wind*, Lawrence Durrell's novels, the classical education of the Great Public Schools, *Zorba the Greek*, a run of BBC serials – the Sons of Zorba, so to speak – and a phalanx of Nordic flesh encumbering remote Aegean beaches. The coral sands and nodding palms are in R. M. Ballantyne's *Coral Island*, R. L. Stephenson's tales, *Bali Hai*, and hundreds of B-grade movies.

The dream of the tropics is in Henri Rousseau, Conrad, Henry Fauconnier's *The Soul of Malaya*. This is not the tropics of Indonesians, of Malays, any more than the dream Mediterranean is that of real Greeks and Italians and Maltese and Turks. They are exotic northern myths, and that is the mood that pervades William Guilfoyle's remarkable creation at the Royal Botanic Gardens at South Yarra in Melbourne. It is not, of course, jungle. Primary or virgin rainforest is fairly open and easy to move through; it is *jungle edge*, as seen from a clearing, or better, from a river. It is there that the foliage is banked in great tapestried walls, dense, impenetrable, green, shining, infinitely varied in texture. It is rainforest seen from the Tweed River, on the border of New South Wales and Queensland, where the Guilfoyles had begun to establish themselves a nursery, and the sugar cane that was later to supplant the rainforest.

Guilfoyle also travelled as botanist to the South Sea island cruise of the HMS *Challenger*, so he had extensive first-hand knowledge of rainforest. That is the image that remained in Guilfoyle's mind's eye, I believe. The point may be established in a mundane way by looking at the main structural elements in the Gardens. There is a total of about 600 palms; 41 specimens of Moreton Bay fig with 110 figs in all; 192 lillypillies; 18 camphor laurels, and 164 southern conifers (hoop pines,

*Adapted from an essay first published as 'About rainforest', *Meanjin*, no. 4, 1990, pp. 700–22.

kauris and the like): but there are only 75 English elms in the Royal Botanic Gardens; oaks total 100 in all, and the true pines (*Pinus*) of all species, 81.

Guilfoyle did more than plant tropical trees. He consciously set out to create a tropical experience throughout the major viewing axis of the site; that is, from the Eastern Lawn across the Central Lake to Princes Lawn beyond, with Government House rising above a luxuriously forested horizon; and the same view in reverse, looking east from the upper boundaries of Princes Lawn. His intentions are described in his annual report for 1875:

> Before commencing to obliterate these walks, I began to form groups in which tropical and subtropical plants will eventually be the prominent features. Through these, dispersed with a view to landscape effect, glimpses will be afforded of the clear lake, studded with islands, the careful plantation of which will materially add to the diversity and charm of the landscape. On this lawn I am endeavouring to imitate as much as possible *tropical scenery* [my italics] . . . It should always I think be the aim of those in charge of public gardens not to reproduce vegetation which may be seen in other portions of such gardens – but to bring before the public, in special spots, scenes of beauty not to be found elsewhere, by representing plants of a different character to those more or less common to the locality. (1876)

The intentions are brilliantly realised, but in a most individual way, for the garden bears little relation to any real tropical environment. It is rather a northern European conception of what the tropics *should* be like.

In my view, neither the greatness and originality of Guilfoyle's achievement nor its costs have been adequately recognised in Australia. He is generally described as a man who worked, albeit in the colonies and in the second half of the nineteenth century, in the great landscape tradition of eighteenth-century England, but this is at best half true, and it obscures his originality. Without question, he handled space – vista, water, reflection, mass, bulk, glimpse, and again, vista – in that tradition. But he was intensely romantic in his intentions, almost to the point of surrealism. The rich, the exotic, the contrasts in colour, texture and form, from the spiky squat-trunked palms he loved to the dense velvet-green backgrounds he gave them, are quite remote from the Age of Reason. He created a dream world from the resources of a unique sensibility. To say that Guilfoyle was perhaps too fond of palms is to misunderstand Guilfoyle.

His most important contribution was his creation of this idealised *tropical* scene, jungle edge as seen from a clearing or river, that Guilfoyle sought to reproduce in South Yarra. The serious student of landscape needs to understand a good deal about rainforest and its plants if he or she is to understand Guilfoyle and his garden. The tendency in Australia to see Guilfoyle as a latterday Capability Brown rather than as the late Victorian that he in fact was, has obscured his originality, yet he is nevertheless also representative of his age and of our Victorian forebears, who were both attracted and repelled by rainforest, for the obsession with rainforest species is not peculiar to Guilfoyle or Melbourne (although it could be said that Guilfoyle encouraged Melbourne, and, through it, other Australian cities, to dream the wrong dream, indeed a wet dream). Late nineteenth-century planting of rainforest species is equally apparent in the Botanic Gardens in Sydney, Brisbane, and, to a lesser degree, Adelaide. Indeed, as in other matters, our Victorian grandparents had a double standard about the trees of the rainforest; they planted them

91

in their parks and gardens much more widely than we do today, especially the evergreen figs, cabbage palms, the *Araucaria* species, *Lophostemon conferta* (the brush box) and *Melia azedarach* (the white cedar, cape lilac or Indian bead tree, as it is variously known), and the camphor laurel, not now indigenous to Australia, though it once was. Even in summer-dry Perth, trees of the eastern rainforest, such as the brush box and cape lilac, along with various lillypillies and evergreen figs, were much planted as street trees. This reflects a set of cultural attitudes in the last century that have not yet been much studied.

Although our forebears liked the trees of the rainforest 'captive', the rainforests themselves were savagely exploited, and, so far as they appear in verse and prose, were seen as satanic. The great, spreading roots of the Moreton Bay fig, for example, become sombre and reptilian; in Christopher Brennan, rainforest becomes a landscape of evil: indeed, 'the heart of darkness':

> the savage realm begins, of lonely dread, black branches from the
>   fetid marish bred
> that lurks to trap the loyal careless foot, and gaping trunks protrude a snaky
>   root,
> o'er slinking paths
>
> (Brennan, 1913 verse III)

Something of this attitude of dread persisted into our own century; when Judith Wright went to live at Tamborine Mountain she became one of the great fighters for rainforest, but it is still rendered as a penitential landscape:

> To reach the pool you must go through the rainforest –
> through the bewildering midsummer of darkness
> lit with ancient fern,
> laced with poison and thorn.
> You must go by the way he went – the way of the bleeding
> hands and feet, the blood on the stones like flowers,
> under the hooded flowers
> that fall on the stones like blood.'
>
> (Wright, 1953, p. 59)

The details are literal, as Brian Elliot points out (1967, p. 320): the darkness at midsummer; the fern; the poisonous giant nettle trees (*Dendrocnide excelsa*). The blood on the stones is the scarlet flowers of species such as *Brachychiton acerifolius* and *Castanospermum australe*, unseen until they fall from the canopy above the forest floor. The precision renders the beauty, but it is still full of menace.

It took Australians a hundred years to learn to love a sunburnt country, and perhaps that struggle made it too hard for us to encompass its contrary at the same time. If we used some trees from rainforest to grace our streets, they were drained of menace by their new context, safely suburban. Dorothea Mackellar includes in *My Country* the 'green tangle of the forest where the lithe lianas coil', but these are not the lines people remember. What is endlessly celebrated in Australian writing is the open, rolling, unbounded, clear, free landscape of pastoral Australia. Closed forest, the technical term which encompasses rainforest, belongs to a different imaginative world.

Another response to the rainforest of the wet tropics is that of Jeannie Baker (1988): she worked in collage, taking living material from the forest. Between Cairns and Cooktown, fringing the mountainous coast of North Queensland, there is a strip of rainforest coming right down to white, crescentic beaches cradled between headland bluffs. It is known to most Australians as 'the Daintree', because it begins north of the Daintree River, crossed by ferry, guarded by crocodiles, but not well enough to stop the bulldozer intrusion, to make a road that no one but the 'developers' wanted. Jeannie Baker worked in this area, and in 1988 I accepted an invitation to launch her collage of six adjoining panels, and a children's book that it was used to illustrate.

The accuracy of Baker's portrayal is adequate to illustrate the structure and ecology of rainforest as I had come to see it through the eyes of mentors such as Dr Len Webb, from CSIRO, one of Australia's experts in the botanical sciences of the wet tropics; also with Colin Totterdell, one of the very best black and white photographers of the natural environment in the world, the equal of Ansel Adams. First, Baker's collage shows a great diversity in a small area. Almost every tree trunk is different from its neighbours, illustrating the great species richness of tropical rainforest, one of its primary characteristics. Subtropical rainforest has fewer species, and temperate rainforest is often dominated by one or two species, for example the groves of lillypilly (*Acmena smithii*) at Wilsons Promontory and elsewhere in Eastern Victoria, and the forests of myrtle beech (*Nothofagus cunninghamii*) in the Otways and Tasmania.

In Baker's collage there is variety in stilt roots and buttresses at the base of the trees. Tropical soils are generally shallow and of limited fertility. The heavy rain leaches out soil nutrients and it also reduces aeration. Because the soils are so shallow, most tropical rainforest trees are shallow rooting and the buttress and plank roots may have some support function, but they are generally also considered to function in gas exchange, like the 'knees' of the swamp cypress (*Taxodium distichum*) in Florida, which also grows in poorly oxygenated soils. Under conditions of tropical weathering, organic matter breaks down very quickly, so the forest floor is almost bare, in marked contrast with, for example, the deep and rich leaf litter on the floor of mountain ash forests in the Dandenongs in Victoria. Most of the nutrients in a tropical rainforest ecosystem are in fact locked up in the living plants (the biomass) and recycled very quickly as trees and leaves die, fall and decay. It is for this reason that the apparent richness of the rainforest is deceptive. Once cleared, tropical soils lose fertility rapidly, as many 'developing' countries in the tropics now find to their cost.

93

The fidelity of Jeannie Baker's portrayal allows genera to be identified, even when only a part of the plant can be seen. The fan palms are *Licuala*; the climbing palms are *Calamus*, the lawyer vine or 'wait-a-while', with its fierce hooks. There are several epiphytic figs, probably *Ficus defluens*. The extensive plank buttresses, the narrow sinuous sheets around the base of the tree in her fifth panel, suggest the Cairns pencil cedar, *Palaqium galatoxylum*, while the trees overhanging the clear creek look like *Tristania*. And so on.

There are several different examples of cauliflory, the name used to describe the ability of certain trees to bear their flowers and fruits directly on the trunk. This is characteristic of some rainforest species, especially some figs and lillypillies. The trees that behave like this usually have flowers that are pollinated by bats, butterflies or moths that prefer the cool and humid world within the rainforest to the much harsher external environment of the canopy. Other features that reflect this constant moist, shaded environment are also in evidence in the collages: lichens, mosses and other epiphytes on the trunks; lianes in great variety scramble for the light, including strangler figs, which begin as creepers, but in time take over their host and become self-supporting.

Accuracy of observation is also shown in the fauna represented in the collage. In the extreme right panel there is a large python, which looks like the amethystine python, *Liasia amethystinus*, restricted to north-eastern Queensland. One would need to get closer to be sure of the identification. But the bird with the brilliant blue wings in the panel second from the left is unmistakable. It is without any doubt, the white-tailed kingfisher, *Tanysiptera sylvia*. It is 'distinctive and beautiful: the only Australian landbird with *two elongated stiff white tail plumes*', says a normally matter-of-fact ornithologist (Pizzey, 1980, p. 216). This bird is limited to tropical rainforest, and its iridescent splendour seems emblematic of the extravagant luxuriance of that environment. It is like the fabled Birds of Paradise in New Guinea, one of which is used on the flag of that country.

The effect is a little like the rather old-fashioned diorama of birds and other fauna in their natural setting constructed by museums the world over – except that this one has a magic, enchanted air. Is Jeannie Baker therefore a scientist, who wishes to instruct us? Not so. She has a message, certainly, which is to communicate her sense of the beauty and luminous calm of that distinctive and threatened environment, and thus to urge its conservation. The children's book *Where the Forest Meets the Sea*, for which the works of collage were made, makes this theme graphically explicit, by superimposing on the pristine environment the outlines (or Mirages) of the kind of tourist developments that are likely to overwhelm it.

But she does not know the scientific names and terms that I have used. Her knowledge is of a different kind, the intimacy that came from living and working in the rainforest for some months, helped to understand it by the family she stayed with, whose knowledge was 'bush-lore' rather than science. And, of course, her work is only superficially realistic. It might be better to call it hyper-real, because it is 'realer than real'. How? First, in the perfect clarity. All the detail is equally sharp, whereas in a real rainforest, the light is uneven, filtered, with some deep shadow and some sharp rays of light. Next, in the physical point of view. The only way that anyone could enjoy such a panorama would be to clear a strip with a bulldozer, and then observe from the cleared strip, yet the feeling of the work is that of being inside the forest, not of observing it from the outside. The view from a road does not give that feeling of being inside the forest, because rainforest seals itself off from such a clearing with a thick mantle of lianes. A photographer can never have a view like hers. When I was with Colin Totterdell photographing the Daintree for CSIRO, I well remember his intense frustration. There was always something else in the way of everything he wanted to photograph. He could never get a long view and there was never enough light, so he was generally reduced to photographing detail, such as cauliflorous trunks, or knotted lianes, or epiphytic ferns.

But she has done more than choose a point of view: she has *composed* a landscape that is harmonious and symmetrical. The symmetry is bilateral but it is not geometrically regular. A small stream creates the focal point, a little off-centre to the left, and the trunks of the trees meet over the stream in the form of a Gothic arch (which carries the idea of a natural church or cathedral). The tree canopy is

Rainforest: the Daintree, Far North Queensland: a collage by Jeannie Baker, consisting of six adjoining panels, with a naked boy in the first (top) panel and a python coiled around a tree trunk in the last (bottom) panel.
JEANNIE BAKER

excluded from the composition, while the vertical trunks reinforce the idea of temple columns. Even the form of the collage itself, a double tryptich, recalls ecclesiastical art and its values ('Forest as cathedral' is one of the recurring images of Western art and literature: Simon Schama (1995) devotes chapters to tracing its lineage and that of 'il bosco sacro').

That Jeannie Baker used the actual materials of the rainforest does nothing to increase its verisimilitude. In fact, that is a red herring. Her technique is remarkable, but its achievement is not representational, although the work *seems* startlingly real, and most people begin by reacting in these terms, as I did. The essence of her communication is not the representation of rainforest, but her experience of it. For her, it was tranquil, happy, innocent, safe. I asked her about these words, which are mine, and she endorsed them strongly, emphasising especially that she felt extraordinarily secure in the forest, although she was there on her own for many hours at a time. This is a very personal vision. Perhaps, like Beethoven, she prefers trees to people.

Although I can give only my own reading of each of these representations of rainforest, surely most people brought up in the culture of the West would bring to their reading images of the 'enchanted forest', which can be malign or benign, or both together. The literary response to rainforest cannot escape the history of Western culture and its accumulated imagery. Whether or not it 'works' for readers depends on the way in which the fresh, the particular, the individual, is linked to the associative resonances of historical experience. Jeannie Baker and her visual imagery may seem to travel light compared with Judith Wright, but the bright innocence of her enchanted forest is also inevitably part of a more general iconography, which includes the world of Henri Rousseau, the brothers Grimm and other fabulists. Her neo-romantic attempt to collapse opposites into each other, rediscovering the garden (nature subdued to human control) within the forest (nature untouched) has its parallels in Guilfoyle and Rousseau. Guilfoyle does this physically, and Rousseau does it graphically, although there is still a hint of wild menace in his seemingly innocent forest-garden. Jeannie Baker's work carries more information about the detail of tropical rainforest than that of Judith Wright, but Wright's responses say more of the human condition.

The scientist seems to have the most direct gaze, looking at the thing in itself, rather than seeing one thing by another, the way of simile and metaphor. Yet the scientist's vocabulary is generalised. There is little profit to be had in arguing that the scientific account is 'objective', or value-free. Luckily enough, the language of the ecologists is full of value terms. None of them can talk of tropical rainforest without using phrases like 'the complexity of the ecosystem', 'the richness in species' and so on. Such words all attribute value to rainforest. The distinctive feature of the scientific account is not that it is objective, but that it is functional. There is an attempt to show how rainforest 'works', and how visually distinctive features, such as cauliflory or plank buttresses, function in this environment, and this gives it meaning of a different kind. All of these discourses have value for me, and I could not do without the scientific.

This discussion leads me to no single conclusion, but it can be used to clarify some areas of confusion. One is a linguistic confusion peculiar to the English language, so far as I know, but increasingly imposed by it on the rest of the world.

It derives from the use of the word 'landscape' as if 'landscapes' had the same kind of physical reality as 'land' or 'area' or 'ecotope'. In Italian, as in French, there is a fairly clear distinction between 'territorio', a piece of land, an area of countryside (without quite that overtone of *possession* that the word carries in English) and 'paesaggio', which is a way of looking at that piece of land. This was the way the word 'landscape' came into the English language, from the Dutch 'landskip', which was a painter's term, and it retained that sense of detached aesthetic contemplation throughout the eighteenth century and well into the nineteenth. The change in meaning, and the loss of clarity, came from North America, where Olmsted and Vaux introduced the term 'landscape architect', and Harvard established the first chair of landscape architecture in 1900.

Now professional landscape architects claim to design and plan landscapes. But no landscape architect can design how I see. They can change landform and vegetation, but how I see it – the landscape – depends on me, not on him or her. This may seem a nit-picking point, but I believe that it is important in the education of 'landscape' professionals, because to ignore the cultural dimensions of landscape may mean that they unwittingly impose their own cultural attitudes on others in the mistaken belief that they are creating objectively good landscapes. Indeed, I have seen this happen often enough, especially in Third World countries dominated by Western professionals, and even in the USA itself, for instance in the resort hotels of Maui in the Hawaiian Islands, which ignore the sensibilities of the Polynesians. My account of Jeannie Baker's work and the contrasting ecological description of the same area illustrate the distinction between 'territorio' and 'paesaggio' rather well.

The examples given may illustrate another point. For some years I was a member of the Conservation Committee of the Australian Academy of Science, and both there and elsewhere, I often heard eminent scientists say things like: 'We must get away from all this emotion, and put our criteria for setting up conservation areas onto a scientific footing'. Although I support this view strongly, it cannot be the whole truth. If we turn for the last time to Jeannie Baker's collage, we may see that in the first panel there is the figure of a naked boy. Why naked? The question is simple, but the answer requires an understanding of some two thousand years and more of European and Jewish history. The forest is harmonious and innocent; the naked human being is young and innocent; there is a serpent in the right hand panel, but even the serpent is innocent, non-venomous, harmless to man, and indeed sometimes kept around the house as a pet and rat-catcher in North Queensland, better than a cat. Clearly we are in the Garden of Eden, or in a version of it, a tropical Paradise. This is the 'good forest', the 'enchanted wood'. Even the bird seems ready to help the boy Adam, like the bird in *Peter and the Wolf*.

Judith Wright describes much the same rainforest, but a different *landscape*. The history of the forest in the European mind is complex, and it includes the dark and dangerous boreal forest of wild and savage animals, the bear and the wolf, the forest of the brothers Grimm and the wicked witches, perhaps a memory of the Stone Age past, but also one that was real enough until well into the Middle Ages and later. The luminous forest of Jeannie Baker takes part of its meaning from the dark and menacing forest of Judith Wright, and we need to understand both if we are to comprehend the complexities of human behaviour in relation to conservation issues. Environmental science is an essential base for environmental planning, but

97

we do not live in a scientific world, and our deepest attitudes, like our attitude to forest, are nearly always ambiguous.

Those who are hostile to conservation may be motivated by perceived self-interest or sheer philistinism or ignorance, but there are also many people who feel insecure in, almost threatened by, the natural environment, and such fears have a long cultural history. Those who support conservation need to recognise this and seek ways of assuaging such fears. Jeannie Baker does it in her collage. Her little boy is visibly *at home* in the rainforest.

Postscript: But this, too, is home

Once, climbing in the Dolomites with an Italian friend, Giorgio Scaggiante, I made a comment on the predatory habits of many Italians in the mountains; they pick or uproot the flowers, herbs, wild asparagus (*cumo*), berries, and shoot the birds, hares, deer, chamois if they can get away with it, all to take back *a casa*. The trip to the mountains is not sanctified unless trophies are carried home. Giorgio was sympathetic, as I knew he would be. 'I say to them', he said, throwing his arms wide, 'but this, too, is home'. Learning to be at home is one of the themes of these essays; living here as if we planned to stay.

References

Baker, Jeannie, 1988, *Where the Forest Meets the Sea*, Sydney & London: Julia MacRae Books.

Brennan, Christopher J., 1913, 'The Quest of Silence', *Poems*, Sydney: G. B. Philip & Son.

Cain, A. J., 1958, 'Logic and Memory in Linnaeus' System of Taxonomy', *Proceedings of the Linnean Society of London*, no. 169, pp. 144–63.

Catalogue of Plants Under Cultivation in the Melbourne Botanic Gardens (1883).

Elliot, B., 1967, *The Landscape of Australian Poetry*, Melbourne: F. W. Cheshire.

Guilfoyle, W. R., 1876, *Annual Report of the Director of the Botanic and Domains Gardens, Melbourne* (1875), Melbourne: Government Printer.

Pizzey, Graham, 1980, *A Field Guide to the Birds of Australia*, Sydney: Collins.

Schama, Simon, 1995, *Landscape and Memory*, London: Harper Collins.

Webb, L. J., and Tracey, J. G., 1980, 'Australian Rainforests: Patterns and Change', in A. Keast (ed.), *Ecological Biogeography of Australia*, 2nd edn, vol. 1, The Hague: W. Junk.

Wright, Judith, 1953, *The Gateway*, Sydney: Angus & Robertson.

10 Home thoughts from abroad: or south-east from north-west

I quoted A. J. Cain in Chapter 7: 'there is no better method for scientists of one period to bring to light their own unconscious, or at least undiscussed, presuppositions . . . than to study their own subject in a different period' (Cain, 1958, p. 44). What Cain says of differences in time is applicable also to differences in place. Half-way down the coast of Italy on the right hand side stands the Adriatic port and city of Ancona, facing Croatia, Albania and Greece. The city was founded by the Greeks. The name is Greek – it means 'elbow', and the map shows why. Australia has two elbows, one at the tip of Cape Range on the west coast, where the coastline reverses direction from north-west to north-east, the other one near Fraser Island, a mirror image. To make Australia, take a rectangular cardboard box, press from above until it buckles out at both sides, then nibble and bite at the top and bottom.

Melbourne lies by this chewed south coast near its lower right corner. North West Cape at the end of Cape Range is the western elbow, and they are further apart than London and Istanbul, and as different. To meet a deadline, I found myself reading and reviewing a book on Melbourne while camping with my son at Cape Range. It was an odd but instructive experience in perspective. John Brack's serio-comic image of Collins Street (Melbourne) at five o'clock shows serried ranks of neat men and women in a grey space, moved along by an invisible sheepdog. Cape Range beggars description, made up of such unrelated elements. It is arid, a dry spine of Tertiary limestone thrusting north-west into the Indian Ocean. The south-west side, facing the open sea, is a national park, which includes a marine park, the Ningaloo Reef, a fringing reef as exotically beautiful as the Great Barrier Reef, not nearly so well known, but much more accessible – once you are there, at Coral Bay or the national park. Getting there is the problem.

Being limestone, the Range is full of caves, and the caves are populated by a fauna of remarkable scientific interest; many species appear to be survivors from an earlier, more humid climate from a very remote past, more remote than the Gondwanan linkages. These ancient troglodytes cohabit the Range with rangers and Range Rovers. Suburban Perth drives up here in July and August to get warm and dry. You can fly into Exmouth, but most holiday makers drive, because you need a vehicle when you get there. The park 'campgrounds' are named and numbered, and the capacity of each is indicated and regulated. You may not camp except at these sites, and when they are full, they are full. They are compact, to save infrastructure costs and maintenance, and generally bare. There are few trees and no privacy. The trek to the toilet is a public spectacle, and needs to be at peak hour in the mornings, to calculate vacancy. 'Camping' almost universally means parking your campervan and living in it. Knowing no better, my son and I put up tents and cooked and ate outside. At first we felt superior: we were *really* camping. Then we felt foolish, physically and socially exposed. The wind blew, the sand blew, the spirit stove blew out, the tents blew down and, one night, came close to blowing away.

It is a bare and open landscape, but you can walk for miles along the coast and rarely see a soul. The water is limpid, turquoise, the beach of fine and very white coral sand; the coast is indented with low limestone headlands and shallow sandy bays. There are fishing eagles and almost no people away from the concentrating campgrounds. Once I met two young Swiss along the deserted beach. I asked them where they came from. 'Lugano' – surely one of the world's loveliest landscapes. I asked them if they were enjoying this bleak coast. They were, very much. I asked them why. The boy extended both arms and smiled. Space.

It is a Space Station; North West Cape has one of the American tracking stations, and the town of Exmouth, on the other side of the spine, was built for it in 1967. Telecommunications still dominate the town, which is not very different from the new suburbs of Perth. Troglodytes to telecommunications gives something of the span of this environment. If it beggars description, it is partly because it belongs to such different time zones, and one cannot therefore read the landscape continuously. The land seems ancient, but it is not so geologically. Cape Range is a Tertiary anti-cline, and both the rocks and the structural geology are much younger than the Precambrian Shield of the south-west and the goldfields.

The anticline brought the petroleum geologists, and nearby Rough Range was the site of the first oil strike in Australia; ironic, because neither it nor Cape Range proved to be productive. The oil show, however, led to further exploration at Barrow Island and the North-west Shelf. It also stimulated an infrastructure of roads and other service facilities that was reinforced by the tracking station. The two roads into Cape Range National Park from the Exmouth Gulf side that display the gorges cut deep into this limestone spine were begun by the oil company.

There are few places in the world so dependent on high technology. This is true of much of contemporary Australia, but brought at Cape Range into sharp focus. Even the campervansters in the national park bring all their food, their fuel and their water – and it is a long drive back on a corrugated dirt road to restock. The park is in fact destocked, in another sense; it was created by withdrawing pastoral leases. 'Ningaloo' is a station name (and before that, an Aboriginal one).

So the landscape is inscribed in diverse hands, sometimes as palimpsest or marginal scribbling, more often as disjunct superimposition. Of Aboriginal land-use there is little obvious evidence, unlike the grasslands and sclerophyll woodlands of a more humid Australia. The sheep are gone from the national park, and the kangaroos have come back in numbers, grazing in sight of the campgrounds, standing at attention by the side of the road at dusk, aroused by the headlights and too often then making the wrong leap. A century of pastoral occupation at very low density is still recorded on the map, and the road from Carnarvon north drives through the leases and their grids.

Fishing and prawning have had a longer history. It is now big business on the Gulf, dominated by the Kailis clan, a Greek family from Castellorizzo, a small and now almost depopulated island in the Aegean. The human history of all this sparsely inhabited coastline from Shark Bay to Broome is nonetheless complex. Pearling at Broome brought Macassans and Japanese divers, Chinese, Melanesians, and small but polyglot fishing communities reached at least as far south as Denham on Shark Bay. There may even be some Dutch blood. There were enough wrecks on the way to Batavia in the seventeenth century, and there is some (medico-genetic) evidence that a group of survivors may have been absorbed into the Aboriginal communities.

Exmouth township is also a tourist resort, but it is not much like Acapulco, Amalfi, Nusa Dua or even Noosa and Port Douglas. There is no pretence at elegance in its caravan parks, and the ambience is more that of the beach shacks and shanties on Crown land that once littered the coast. The caravan parks are gregarious places, friendly, casual, tending to the down at heel. The average age of the customers is high. Many are retired, and many are regulars, driving north every winter. It is a long drive, 1300 kilometres, and in this sense, Cape Range is remote, much more so than Bali or Jakarta, a short flight from Perth, or Singapore, only an hour or so further.

Yet it is not at all a frontier town either: all is plastic, disposable, microwave, fast food – and like the snail and the turtle, the visitors bring their homes and their way of life with them. They drink beer and catch and clean, cook and eat, the fish. The voices are raucous, the smiles ready. The faces of the old salts are lined and the skin parched. There are also a few families with small children, and a sprinkling of cosmopolitan travellers, often young Germans. Japanese honeymooners are not in evidence.

It was here, camping in the adjoining Cape Range National Park, that I read Janet McCalman's study of Public School Melbourne: *Journeyings* (McCalman, 1993). I had no choice, as I was committed both to the holiday with my son and to meeting a *Meanjin* deadline. Her book gives an account of the changing social and physical landscape of the 69 tram that runs from St Kilda along the length of Riversdale Road, and thus serves most of Melbourne's private schools. Cape Range was a bizarre place to read this evocative portrait.

From the viewpoint of the Cape Range and Exmouth Gulf, middle-class Melbourne seems claustrophobic, huddled irrelevantly at the wrong end of the continent, near nowhere but Launceston, cut off from that rapidly growing link to Asia that is so strong in Western Australia, and far less probable as an *Australian* settlement than the transient, arid, high-tech environment of the Cape. The world of the book seems confined by that grey city, and only in a few rare passages does it break out. I cheered for the man who said that Scotch College taught him 'how to read a compass and a map, put on old clothes and disappear into the bush' (p. 283). But apart from one escape to the bush and a few war experiences, we are rarely let out of reach of the number 69 tram, running through the heartland of middle-class Australia (see colour illustration following p. 78).

Yet Melbourne has its own meaning, and is not as 'isolated' as it may seem. It is still home to most of the big mining companies. BHP (Broken Hill Proprietary) is

101

international. Comalco pioneered bauxite mining at Weipa, under very difficult conditions, and many from Head Office, many of them from Scotch College, were full of robust pioneering zest. The excitements and achievements of scientific research, much of it by men and women from the private schools, have made a great contribution to the life of this country. The middle-class have also shown a flair for self-criticism, although surely Barry Humphries is quintessentially a class product (from Melbourne Grammar, in fact – as is John Brack, whose *Collins Street* says so much about the repressive aspects of the middle-class culture).

It is inconceivable that a book such as *Journeyings* could have been written in Perth, and unlikely in Sydney. They are both too pseudo-extrovert and more likely to hide their anxieties and neuroses than to face and explore them. The book is itself a product of that middle-class Melbourne culture of which it is also a thoughtful critique. Jim Cairns and the Vietnam rallies were typically Melbourne, like so much social change generated by moral concern and critical introspection. If its middle class remains the 'anxious class', Melbourne is still also in many ways the heart and mind of Australia. But then I have written the last two paragraphs from the viewpoint of a recollected Melbourne. The point of the contrast is that it looks very different from the Cape. Each place has its own truth. Generalisations about Australia and about places are contextual – a thought that leads directly into the next section, on sense of place.

References

McCalman, Janet, 1993, *Journeyings: The Biography of a Middle-Class Generation 1920–1990*, Melbourne University Press.

Seddon, G., 1993, 'The Anxious Class: An Essay Review of *Journeyings* (by Janet McCalman)', *Meanjin*, vol. 52, no. 3, pp. 582–8.

THEME ⸬ ſ ſ ⸬ ſ   Locating: the sense of place

Chapter 11, Sense of place, the first of the essays in this section, was written in 1971. *Sense of Place: A Response to an Environment* was published in 1972 by the University of Western Australia Press. It has always been my best-known book in Western Australia, and people continue to say kind things about it. On my return after twenty years, people come up to me and say: 'I really like your book', and I have no doubt which one they mean. I am also asked fairly often to speak on '*Sense of Place* revisited', and again I know at once what is intended: 'What physical and social changes have taken place in the intervening years; is our behaviour now better adapted to the physical constraints of the environment, and are our buildings and urban fabric more sensitive to  it, or do they ignore it?'.

These are good questions, and I try to answer them, with a mixed bag of answers – for example, Western Australians are much better at dune conservation, but worse at water conservation (our consumption has increased steadily, and the household average consumption is now 515 kilolitres per year; if this were milk rather than water, you need to imagine the delivery of 515,000 one-litre cartons to your door every single day of the year). Limestone, the bones of the coastal plain, is now popular, too much so, since it is a consumable resource. As to attitude, more than half the population of Perth in 1991 was not here in 1971, and many of those who were here then had arrived as part of the migrant influx, primarily British, in the previous decade. They sought or were sold a 'palm and pool' image of the place, a lush subtropical paradise of hibiscus and immaculate lawns, rather than that of a place with a mild, wet winter but a long, hot and very dry summer.

But a different question, 'Is "sense of place" a valid concept; is it still a relevant one?', is not asked. An affirmative answer is taken for granted. 'Sense of place' has become a popular concept, heard at every turn, unanalysed, and this is, for me, a problem. It is partly the contrariety of the academic that once a concept is popular I tend to lose interest in it. But it is also the belief that this concept should be applied with caution, because it is a form of appropriation. It can be a way of legitimising a set of personal and subjective evaluative criteria as if they had some externally derived authority.

My book *Sense of Place* is in part an old-fashioned regional geography of the kind that England and most of Europe has in plenty. Geography, however, was largely a post-World War II profession in Australia (apart from Griffith Taylor in Sydney), and by the time it arrived in Australia most geographers had turned away from the old integrative role to that of the more narrowly focused scientist, producing papers on 'The morphometric study of slope retreat' and the like – the exception being the tiny handful of historical geographers, who have written some of the best accounts of landscape in Australia.

We still need the integrating perspective, so I do not disown *Sense of Place*. It is the kind of book I should like to have been handed on arriving in Western Australia, to help me orientate (occident?) myself. Note, however, that this is not a question of identifying 'the sense' of 'a place' as if it were already there, waiting, but rather of finding *my* place in a new setting. It was also a tract for the times, and a very necessary one, in that current development showed scant regard for the physical parameters of the Swan Coastal Plain. In retrospect it seems to me to have a certain innocence, as does Chapter 12, *The genius loci and the Australian landscape*, which has been required reading for most courses in Landscape Architecture in Australian universities at one time or another. In both cases, the key terms, 'sense of place' and '*genius loci*', are assumed rather than analysed, although that might be forgiven if one accepts the reason for writing them at the time. Again, I do not disown the term *genius loci*, but urge that it be used with caution. It would be dangerous to assume that there really is a *genius loci*, that our task is to identify it, and that its discovery then justifies the design decisions we choose to make. It is a useful concept, but it is culture-bound, so that its application requires critical self-inspection.

In the version published in *Landscape Australia* in 1979, the illustrations need such scrutiny. A whitewashed timber bridge over the Ovens River is described as 'right for Bright', and I still find it more sympathetic than the concrete bridges that are replacing the wooden ones – but perhaps this is locking Bright into a romanticised past. A 'new' (1970s) farmhouse is illustrated, 'in a timeless Australian style', that is, with the roof-line continued all around the house as a deep verandah. I still think that this house form is well suited to most Australian conditions, but where the form had just begun its revival in the 1970s, it is now endemic, and we are building a whole world of copies. The future appears to lie with the past in some places, notably in Perth, where every second new house is 'Federation' – a fad not without irony, in that Perth, along with most of the State, was carried into Federation, not from its own inclination, but from the strong vote from the gold-fields, which tipped the balance. So we are not only 'future eaters' (see Chapter 24); we also devour the past, now a kind of tourist destination.

Another essay on this subject is 'The Rottnest experience': Rottnest is an island offshore from Perth, very popular for relaxed summer holidays, and increasingly used all the year round for walking and cycling. Although it is now much loved and has a kind of paradisal standing with its devotees, it has had a dark history. I wrote the essay in 1983 for a special issue of the *Journal of the Royal Society of Western*

A whitewashed timber bridge over the Ovens River: 'Right for Bright'? GEORGE SEDDON

Australia: it is also included in a collection of my Western Australian essays under the title *Swan Song* (Seddon, 1995), so it is not included here, but it is relevant to the present theme. A question about 'sense of place' – '*Whose* place?' – has become critical at Rottnest. Aborigines from all over Western Australia were imprisoned at Rottnest for much of the nineteenth century and the early years of the twentieth century, even from the Kimberley, and the deprivation such displacement entailed for people who were so closely tied to their 'country' was acute. The image of Rottnest held by many Perth people, especially of an older generation who spent carefree summers there, is arcadian, a reminder of a better, simpler world (whether it really was either better or simpler is a different question, and one that is explored in the essay). But the Aboriginal 'Rottnest experience' was bitter. Aboriginal graves have been located in the camp ground, and in 1996, an Aboriginal group laid claim to the whole of the island as a major repository of their history, focusing *their* sense of place.

Chapter 13, Cuddlepie and other surrogates (1988), is about the endowing of non-human creatures with a human face; there is a wealth of possible examples, from Mickey Mouse to Winnie the Pooh. May Gibbs' characters, Snugglepot and Cuddlepie and the rest of her 'bush creatures', are a special form of 'environmental education'. They are part of an ancient culture.

Chapter 14, Jet-set and parish pump, was written in 1973. It takes a broader perspective in looking at two opposing cultural tendencies, towards an inter-nationalism that is also homogenising, and towards a regionalism that can be parochial and restricting, creating ethnic prisons. The excesses of both movements are more apparent now than ever before: McDonald's setting up shop in China on the one hand, the horrors of ethnic cleansing in Bosnia on the other. Because the opposing tendencies are still with us, I have added a new essay in which I attempt to put the debate in context and to locate my own place in it; hence Chapter 15, Placing the debate.

References
Seddon, G., 1983, 'The Rottnest Experience', *Journal of the Royal Society of Western Australia*, vol. 66, parts 1 and 2.
Seddon, G., 1995, *Swan Song: Reflections on Perth and Western Australia, 1956–1995*, Centre for the Study of Australian Literature, University of Western Australia.

11 Sense of place*

I grew up in country towns in Victoria. From Mildura, a compact irrigation settlement along the Murray River, we could ride a bicycle for a few hours and camp the night in empty country, or so it seemed then, with nothing but crows and river gums, and the broad Murray for fishing and swimming. Later, we moved to the Wimmera. I went to school in Ballarat, on the edge of town, where the fields began. They were fields, too, rather than paddocks; not very big, often with gorse around them, unwisely brought by early settlers, and double rows of pines as wind breaks in that windy place. It was a formed landscape, with landmarks made by men, women and Nature; and it was a landscape with a past. The natural landmarks were two worn-out volcanic hills, Mt Buninyong and Mt Warrenheip, both with clear outlines that changed as you saw them from different directions, but always distinctive on the skyline.

Behind the school there was a mullock heap, an abandoned pile of mining refuse. Ballarat had made its money from gold in the roaring days eighty years before I got there; what was left after the gold had gone was a solid, well-built town of stone, set on a low remnant of the Great Dividing Range in good farming country. There were also a lot of dead Chinamen in the cemetery, Irish names on the pubs, and mullock heaps. The word 'mullock' was a dialect word in England 150 years ago for rubbish or dust, ultimately from the Teutonic root *mul*, meaning 'to grind', but the word survives now only in Australia and New Zealand. Behind the mullock heap was Dead Horse Creek, a gentle stream despite its name, with a white-railed wooden bridge, and beyond that, a good gravel road with hawthorn hedges along the side of it in places; it went to Mt Rowan, a rounded hill a few hundred feet high and well grassed. I once rode a bicycle from Ballarat home to Horsham through Ararat and Stawell, and another time I rode down to the coast at Portland. One way and another, I got pretty much all over Victoria, but hardly ever out of it, except for hitch-hiking up the Coast Highway to Sydney once, and for a summer walking

*First published as 'Foreword' in *Sense of Place: A Response to an Environment*, Perth: University of Western Australia Press, 1972.

through the Tasmanian highlands and strawberry picking in the Huon Valley. When I was at the University of Melbourne I went walking through the Victorian part of the Australian Alps two or three times a year, and got to know them well. This was before the Snowy scheme (discussed in Chapter 5); there were no roads, and you carried up to two weeks' worth of food. It was a stiff climb up the valleys on to the dissected plateau that makes up most of these highlands, but when you got up on top it was like a golf-course without any golfers. I don't recollect ever running into another party when we were up there.

When I was twenty-two I went to England, and later to Portugal and Canada. I was away for over six years, and I had a good time, but I was often homesick for Australia, for the smell of the bush, and the trout straight out of the Howqua on to the pan, and the springy turf and great skies of the Bogong High Plains. My last view of Australia had been of Perth, the only time I had seen it, on a cloudless day, and I remembered Kings Park and the river below. At the University of Toronto I heard of a job at the University of Western Australia so I applied for it and got it. I had been away from Australia for long enough. I had married in Canada and we set sail. We landed in Sydney, drove across the Nullarbor and got to Perth. I was home in Australia.

I hated it. Partly this was the malaise of the returning traveller, partly cultural snobbery, partly genuine regret for the more cultivated society I had left behind. This story has been told often. But I had a more specific grievance: I just didn't like Western Australia. The country was all wrong, and I felt cheated. This wasn't what I had come back for; where were the ferntree gullies, the high plains, the trout? All the plants scratched your legs. The jarrah was a grotesque parody of a tree, gaunt, misshapen, usually with a few dead limbs, fire-blackened trunk, and barely enough leaves to shade a small ant. If you went camping in the summer, you carried water – you couldn't take a running stream for granted. It slowly became clear to me that I wasn't an Australian at all, but a Victorian. Here I was stuck on the edge of a continent I knew next to nothing about, and didn't like. But I liked the City of Perth itself, and the people, and the job, so we stayed. Slowly I came to understand the land better.

That is the point of the autobiographical introduction. I was ill-prepared for Western Australia, and I think this must be a common experience. Even the Western Australians whose families have been here for three and four generations are ill-prepared in some basic ways, because our primarily British background is still apparent in our attitudes towards the way we use the environment; and in nothing so much as our attitude to water. Centuries of water-riches makes it hard to grasp our water-poverty and its implications, although these are better understood in Western Australia than they are in the coastal cities of eastern Australia. In Perth (and Adelaide) the aridity of this most arid of the continents is a part of one's consciousness. The south-west of Western Australia is an island, with sea to the south and west, and one cannot leave it north or east without crossing mile after mile of desert.

But it wasn't the desert I disliked, it was the country within a hundred kilometres of Perth. The coastal plain is flat: the Darling Ranges are nothing more than a 300-metre step-up to the flat plateau. The vegetation seemed monotonous, the soils little more than sterile sand. And apart from the upper Swan Valley with its vineyards, there was no landscape.

An environment becomes a landscape only when it is so regarded by people, and especially when they begin to shape it in accord with their taste and needs. Nature may offer the raw material of scenery unaided, but to transform it into landscape demands the powers of the seeing human eye and the loving human hand. In much of the country around Perth, the hand of man has been heavy, and my eyes were not able to see the subtleties of the natural landscape.

So my first job was to understand this land. I was lucky in the people I met, and I began to learn. Much of this learning is called 'science', but I would rather forget that word for the time being. For me, it was learning to see. Like so many others, I began with the plants. The wildflowers are easy to like. Then I became interested in the way different plants make a living, and how they have come to be as they are; and so I was off into evolutionary history. In time I came to find this an absorbing place, full of questions and rewards. I must have acquired a new range of landscape images from the painters, although I have no memory of the transition. Sir Arthur Streeton's hazy landscapes, and others of the Heidelberg School, were supplanted in my mind's eye by the gaunt, clear images of Sydney Nolan, one of whose best paintings hung in my study at the university for years. But I also learned to change focus. Much of the interest of the landscape around Perth comes from very small things rather than from 'scenery' as conventionally understood, and this change in the scale of attention is also one of the moves of science. Finally, I became interested in the structure of these landscapes, the evolution of the landforms and of the coastline, of the creeks, rivers, swamps and offshore islands – and of the way in which people have used them. In *Sense of Place* I have put together some of what I have learned about this part of the world by reading and looking. As well as imparting information, I have tried to communicate something of what acquiring it has been for me, the experience of an environment.

In a book about the attempt of poets to come to terms with the Australian scene, Brian Elliott begins thus:

> The first need in a new country or colony must obviously be in one way or another to comprehend the physical environment. In poetry we find this need reflected, in colonial times, in an obsessive preoccupation with landscape and description. At first the urge is merely topographical, to answer the question, what does the place look like? The next is detailed and ecological: how does life arrange itself there? What plants, what animals, what activity, how does man fit in? The next may be moral: how does such a place influence people? And how, in their turn, do the people make their mark upon the place? How have they developed it? Next come subtler enquiries: what spiritual and emotional qualities does such a people develop in such an environment? In what way do the forces of nature impinge upon the imagination? How do aesthetic evaluations grow? How may poetry come to life in such a place as Australia? (Elliott, 1967)

My queries, although not so complete, follow this sequence. The book is in three sections: The Land; The Plants; Man. There should perhaps be a section on the animals other than humans, but I don't know much about animals, mainly because I am too slow-witted to study things that move. Trees and rocks are usually to be found where you left them if you come back in a month or a year, and therefore are more satisfactory from my point of view. So I have discussed wildlife only where it seemed essential to the theme.

The emphasis of the book is environmental design. The word 'conservation' has some negative uses, especially when it is used to mean the same as 'preservation'. To some people, all countryside that has been touched by man is spoiled, and so an unspoiled environment is an empty one. This may be true in particular cases, and it is often necessary to fight for a specific area against the improvers. In a report by the Swan River Conservation Board, an exemplary body in many ways, we are told that the Board 'is still developing the Upper Reaches of the Swan River' (Swan River Conservation Board, 1970, p. 1). The development consists mostly in dredging, desnagging, filling the low-lying areas and planting couch grass. Later we are told that six riverside acres have been bought in Middle Swan by the Swan Shire and will be developed as a picnic spot. '*Even in its natural state* it provides a very attractive rendezvous for a river excursion. Temporary toilets have been erected' (p. 6, my italics). One thinks of the eighteenth century: 'I hope I may die before you', said one of Capability Brown's contemporaries, 'so that I may see Heaven before you improve it'.

But it is not possible to deep-freeze the ecology of a whole area, or go back to the Swan Estuary as it was before the first Kleenex was dropped. The most enchanting dream that has ever consoled mankind is the myth of a Golden Age, in which we lived on the fruits of the earth peacefully, piously, and with primitive simplicity. But the complex problems of today's world are hardly to be solved by our all going camping, or sitting around taking in each other's wishing. The view that an unspoiled environment is one untouched by humans can hardly be pushed to its logical conclusion, and in any case it is misleading, first because it sets Man against Nature, where it is more illuminating to see Man as a part of Nature; and secondly, because we are not always despoilers. We can also be creative. Some of the world's finest landscapes are man-made: much of Europe, for example, or the terraced hillsides of the Philippines, and Machu Picchu in the Andes. These are all examples that were moulded before the technological revolution of the last hundred years, but one could add to the list the graceful concrete rail bridges of Switzerland, each of which is a work of art that ties together the elements of a fine natural landscape. The Narrows Bridge in Perth is both functional and beautiful, and it enhances the landscape of which it is a part. The concrete mixer and the bulldozer can make landscapes as well as mar them, and the emphasis in conservation, especially in urban areas, should be on intelligent land-*use* and on environmental design and not just on preservation, although there is also much that should be preserved. The first step in design is recognition, the ability to see what there is. Only then can we ask whether a given structure is appropriate to its setting, or whether a proposed land-use is appropriate in a given environment.

References
Elliott, B., 1967, *The Landscape of Poetry*, Melbourne: F. W. Cheshire.
Swan River Conservation Board, 1970, *Up-river to Middle Swan*, Perth.

12 The **genius loci** and the Australian landscape*

There is a latter-day proverb that anything xeroxed begins to lose its value. This is the Age of Easy Copying. I have a similar maxim of my own, that any place that you can get to by jet is unlikely to be very different from the place you just left. International technology and easy communications have a homogenising effect all over the global supermarket (whatever else, it is no village), and thus works to reduce regional diversity in urban form, architecture, food, dress and even vegetation. For example, as recently as the late 1950s, most cities in the humid tropics were distinctive at least in their trees. Now there is a very limited and small vocabulary of street trees from Cairns to Chiang Mai: *Delonix regia*, the royal poinciana; several species of *Cassia*; *Spathodea campanulata*, the African tulip tree; a few palms; *Terminalia catappa*, the tropical almond; *Dillenia indica*; and the ubiquitous rain tree. These are all magnificent trees, extravagantly beautiful in flower, but it is extraordinary that a part of the world remarkable for the diversity of its tree species should so quickly be reduced to uniformity in its towns.

113

Does it matter? I believe that it matters a great deal, for a variety of rather different reasons, ranging from ecological through aesthetic to psychological. Some of them will be illustrated later. The psychological dependence on regional and local identity is marked in all technologically 'primitive' societies – in much of Papua New Guinea, for example, there is a wealth of local knowledge about particular places, their physical characteristics, their special uses, and the kind of behaviour that is necessary to maintain both, encoded in a series of taboos, negative and positive constraints on behaviour that Westerners have sometimes described as superstition. The early Greeks had very similar beliefs and practices, and a legion of tutelary deities – the spirits of place – to guard each stream, grove and mountain. That body of legend embroiders the cloth of our Western tradition, but it now has no functional place in it, only a decorative one.

We make sense of experience by generalising, and could not function without so doing. Every noun in our language takes meaning from a perceived likeness

*First published as 'The *genius loci* and the Australian landscape', *Landscape Australia*, vol. 2, 1979, pp. 66–73.

between different objects, and every verb, from a perceived likeness between different events or actions. Yet experience itself is of the specific, and each of us is an individual with a need to see ourselves in a unique set of relations, as well as in general ones. This need is not fully met in a homogenising world, and many aspects of our life, including much of the fashion industry and the featurism (detail that has no function other than to distinguish one house or garden from the next) still characteristic of domestic architecture and suburban gardens are evidence of unfulfilled cravings for personal identity set in a distinctive environment.

Those who are responsible for the care of landscape in this country can do much to resist the effects of homogenising technology, to individuate by understanding and clarifying the locally distinctive – in short, by respecting the *genius loci*. What follows are some suggestions as to how this might be done, and some possible reasons for doing so. The suggestions are all commonplace, but they are nevertheless regularly disregarded.

Learn the geology

Understand the geology, and display it where you can. Road and rail cuttings often reveal the geology of an area in a satisfying way. If the rock is not strong enough to stand as a cutting, then it may be necessary to face it, or to use crib walling or reduce the slope and cover it with turf. But it is not uncommon for the natural rock face to be covered over for reasons that are primarily aesthetic. In Melbourne, for example, one of the most significant internal boundaries is that between the Tertiary basalts of the western third of the metropolis, and the Silurian mudstones and sandstones of the eastern two-thirds. This boundary coincides in part with the Yarra, but it is not the river that makes the significant boundary, but the geology, as Ivanhoe and Heidelberg show clearly (both 'eastern' suburbs, west of the Yarra, but east of the basalt). This transition is shown well in Barkers Road in Kew and Hawthorn – the road leaves the basalt flats of Collingwood to cut through the Silurian rocks of the high eastern ridge along the river. Further south, Burwood Road makes a similar entry, but here the cutting has been 'beautified' with basalt terracing, planted out with gazania, agapanthus and other pleasant but commonplace exotics. It is also very fashionable in the eastern suburbs of Melbourne to use large basaltic boulders to make gardens in imitation of the Ellis Stones gardens. Basalt does not belong in these areas, and anyone who uses it should be very clear in his own mind just why it is being introduced. Fantasy gardens that create their own environment are legitimate in special circumstances, especially if they are self-contained and not continuous with a larger, semi-natural environment, but they should be exceptional.

The gold of Hawkesbury Sandstone is part of the riches of Sydney, and it is everywhere, in cliffs, cuttings and buildings. Rock and stone are as a rule used well in Sydney, because the natural example is omnipresent, yet even Sydney has some very inept stonework; along sections of Mona Vale Road constructed in the late 1970s, for example, cuttings are faced with a veneer of crazy paving. The exposed surfaces represent bedding planes, but are laid at a 30 degree angle to the true bedding planes of the Hawkesbury Sandstone. Toodyay Stone, a hard flaggy quartzite, is often used thus in Perth and Fremantle, with deplorable results. The outcropping limestone (calcareous sandstone) of coastal south-western Australia has a distinctive beauty that seems not yet to be adequately valued, since outcrops are often vandalised by developers and public authorities.

Respect the landform

Study the landform, and build in sympathy with it, if possible. The handling of land-form is particularly clumsy and insensitive in Australia, although many farm-houses of the last century were beautifully sited. Camden Park House, built by John Macarthur's sons, is a striking example, facing west down a long gentle incline to the Nepean, rising again to a low hill where, a little off-centre, stands the tall tower and spire of Camden's Church of St John, built by the Macarthurs to the Glory of God and to improve their views; behind it in the distance, the backdrop of the Blue Mountains with the peaks of Mt Victoria and Mt Wilson accenting the skyline. This view is a combination of nature and artifice; the house is angled for it, the unbroken lawn leads to it, the flanking tree plantations frame it, the church on its hill enhances the middle distance, the Blue Mountains close it. The view from the opposite side, the east or garden front, complements it by being so different; the garden is set on a rather small plateau or shelf of land, planted as beds around a semi-circle of lawn dropping steeply to broken country, a billabong, and natural bushland. It is more civilised in near view, much wilder in the background, and all more intimate in scale, than the sweeping vista to the west on the other side of the house.

This eye for a site was not limited to great houses. It was as true for many a modest farmhouse, but it is a skill that has been lost. A 'view' today seems to mean only a large range of visibility, as from the West Gate Bridge, or from aggressive houses rearing above the landscape on the Mornington Peninsula or Palm Beach to peer at equally aggressive neighbours. Skillion roofs are set at an angle contrary to the slope; the flow of the landform is disrupted by trivial shapes, or by brutal cut-and-fill to produce level building sites. Most of the Yarra River has been filled along its banks, flattening out the swamps and billabongs, and oversteepening the banks to leave the river out of sight in a trench with steep and unnatural banks. Landscape architects have a little trick all their own called 'mounding'. The creation of new landforms with big earth-moving machines has exciting possibilities, of which the underground car park at the University of Melbourne gives an interesting example. It is also sometimes convenient to use mounds of earth to screen or to reduce noise. But the arbitrary tumuli or long barrows that have appeared in affluent suburbia – the long barrows of Mona Vale in the northern suburbs of Sydney are a major example – have nothing to do with natural landforms on this continent, although they bear some relation to glacial landforms in North America and Europe, whence the device has been copied. I hope also that we will have no more waterfalls beginning and ending nowhere – even the waterfall in Commonwealth Park in Canberra, on a generous scale and exceptionally well done, is ludicrous to my eye. No natural spring could possibly emerge from that granite-bouldered construct on the limestone plains by the Molonglo.

Some public agencies showed a new feeling for landform in the 1970s. In Victoria, the Country Roads Board created a fine megasculpture in the Wallan bypass section of the Hume Freeway; the highway is beautifully fitted to the land in long sweeping curves. The overpasses are tied securely to the hills from which they flow, and the road carries the traveller up and over the Dividing Range with kinaesthetic pleasure. The State Electricity Commission (Victoria) took great pains in the difficult exercise of siting its power lines in the 1970s and 1980s. Fisheries and Wildlife had some interesting design successes, most notably the information

115

building at Tower Hill near Warrnambool. This was designed by the late Robin Boyd, and the circular building with domed roof faithfully reflects the bare rounded hill forms behind and around it. It is unusual for a state agency to sponsor work of such distinction.

Respect the soil

Study the soil. Any gardener will improve the soil if he can, by digging in compost, adding gypsum to clay, and by fertilising. For an enclosed garden, it may even be legitimate to replace the soil entirely, with 'mountain soil' or an organic-rich sandy loam. Remember, however, that you are certainly robbing Peter to do so. In the larger landscape, excessive fertilising will mean that there is then a temptation to use plant materials that would not otherwise grow on site, and this may commit you to substantial long-term maintenance. There may also be ecological costs, and aesthetic ones. The fertilisers will add to the eutrophication risk to streams and lakes. Changes in vegetation may make a landscape more vulnerable to adverse conditions, reduce wildlife habitat, and look out of place against the bleached colour typical of so many Australian landscapes.

Respect the hydrology

Interfere as little as possible with the natural hydrology. To increase or decrease stream flow will lead to either erosion or siltation. Where the velocity of run-off is necessarily increased in one part of the system, for example by roads, this should be compensated by decreasing run-off elsewhere, for example by infiltration beds, reforestation, or retarding basins. There are now many devices for the systematic analysis and conservation of natural hydrological systems, and these give a better long-term return on investment than the very costly corrective devices required by ignoring them, a case where 'Design with Nature' is good practical advice. There are also some obvious aesthetic rewards for doing so.

Respect the natural vegetation

Study the natural vegetation, and the existing vegetation. The design of self-maintaining systems should be a general aim, although it can rarely be fully achieved, and there may be departures from it for specific, clearly defined purposes. One must also recognise the identity of cultural landscapes; the poplars of the middle Hawkesbury around Wisemans Ferry; the gentle vineyards of McLaren Vale; the great pine and cypress windbreaks of the rolling lava plains near Flinders in southern Victoria, almost black against summer-blond pastures; the backyard almonds of Adelaide and the elms of inner Melbourne, where a eucalypt may look and be out of place. To grow only native plants is not a proper aim, which should be rather to grow appropriate plants – but the definition of 'appropriate' must rely on much physical and cultural information.

One should nevertheless use local plant material, germinated from a local seed source, in the large landscape as a matter of general principle. Departures from this principle may often be legitimate, but should have an explicit justification. The variation in plant material is one of the strongest natural cues to changes in the local environment, but once again the forces of homogeneity are at work. In the Hume Freeway section praised above, there is one landscape inadequacy, and that is in this use of plant materials. A rather limited range of landscaping plant material has been used, drawn for its practicality and easy availability and familiarity from around Australia. Perhaps one-third comes from south-western Australia. The Dividing Range is one of the great internal boundaries in a country that generally lacks marked differentiation, separating the southern coastal margins from the inland Murray Basin. This major transition is not recognised in the planting scheme on the Hume Freeway, which thus homogenises its route, and diminishes the *genius loci* in this respect.

It sometimes looks as if the rather limited range of imported exotic plants used to adorn our gardens and landscapes in the past has been replaced in the last decade or so by an equally limited range of native plants – native, that is, to Australia, but rarely indigenous to the site where they are planted. I am not an extreme purist, and am well aware, for example, that *Eucalyptus ficifolia* grows much better in Melbourne than it does in most of Western Australia, where its natural range is limited to a small area on the extreme south coast. There is in fact a vast range of Australian plants that has scarcely been tried in general cultivation. Many of the rainforest trees, for example, thrive in the Royal Botanic Gardens in South Yarra, but are almost never to be seen outside them in Victoria – the species of *Lomatia* are a case in point. There is also a very great deal of work to be done in the design use of Australian plant material, a subject which people like Glen Wilson have pioneered (followed more recently by Kath Geery, Bill Molyneux and Diana Snape, among others: see Snape, 1992). The use of repeated verticals, so easily achieved by close-planted species such as *Eucalyptus maculata*, can give stunning effects, and there are many striking design possibilities which rely on the light canopy, fine and diverse foliage, subtle colour variations, sensuous trunks and irregular symmetry peculiar to so much Australian vegetation. All this should be explored for special effects. In the larger landscape, ecological principles should generally apply. They can also be very effective, soothing by their harmony in an urban setting, as Bruce Mackenzie (1979) has shown so well at Peacock Point.

Respect the cultural landscape

Discover the cultural landscape. There are many good local and regional landscape practices: the whitewashed low wooden railing around the country race-track, for example, is preferable to the tubular steel with which it is sometimes replaced. Street names attached to the walls of buildings, rather than to yet another pole rising from cluttered pavements, is an old and urbane practice in inner Sydney. The pepper tree (*Schinus molle*) is as much a part of the cultural landscape of Australian mining towns as any gum tree. The cultural landscape is made up of the sum of such details, reflecting our impact on the natural environment. Unfortunately, this principle cannot be applied incautiously, because much of the cultural landscape is not worth respect, a sad truth that applies not only to the dreary landscapes of part of the western suburbs of Sydney, which are often subject to the scorn of aesthetes, but equally to most of the Barrenjoey Peninsula, where many of the aesthetes live, in an area of great wealth and breathtaking natural beauty among houses of equally breathtaking vulgarity. However, human use has enhanced some landscapes: the Fleurieu Peninsula in South Australia, for example, and much of rural Tasmania. Such landscapes must be conserved, and their lessons applied elsewhere.

117

Context and the **genius loci**

Analyse the *genii loci* of our landscapes and celebrate them. The landscapes of England have been loved, analysed, painted and used as the background for poems and plays and novels for hundreds of years. Wessex is Hardy country, and the lakes are Wordsworth; East Anglia is interpreted through Constable, and Kent through Chaucer. This rich overlay of association deepens our experience and understanding. In Australia, Heysen invented the gum tree, and Tom Roberts showed us reddish landscapes, a change from the sombre colours of McCubbin, or the south-of-France palette of Streeton. But most Australian landscapes are unlimned and unsung. Dorothea Mackellar and her pop poem about loving a sunburnt country (which she did from a house on lush Pittwater with an annual rainfall of around 1300 mm) has probably done more to wean Australians from the hose than any landscape architect; programs like those of the ABC Natural History Film Unit, which does work of world quality, have also had a major impact.

The view of some professional landscape architects that the way to get good landscapes is to get on with the job of design, and that words are all a waste of time, is clearly inadequate. Landscape architects will never play more than a partial role in managing landscape in this (or any other) country, but it can be a critical one if they are articulate, can say what they mean by good design as well as show it, can give meaning to words and phrases like 'compatible' and 'incompatible', 'in harmony with the landscape' and so on, and by the capacity to analyse in words – and photographs and sketches – the specific qualities of specific landscapes, thus showing us all how to pay homage to the *genius loci*. This is a task that has barely begun.

Conclusion

These seven points are commonplace. Any landscape architect is likely to have heard them all in the first few lectures of his first term of professional training. Some of them are put into practice on occasion, but there are few practitioners who make a conscious attempt to apply all of them all the time, with the outcome that some landscape architects are helping to create a Hilton International landscape, a bland, easily digestible setting that makes no demands, belongs to no place, and make the international middle-class travellers feel at home anywhere in the world; the price that they pay is that they are really at home nowhere.

Moreover, much landscape management is in the hands of people other than landscape architects – shire engineers, for example, who may *not* be familiar with these simple principles and may need help in applying them.

References

Mackenzie, Bruce, 1979, 'Peacock Point, Sydney Harbour', *Landscape Australia*, no. 1, pp. 19–27.

Snape, Diana, 1992, *Australian Native Gardens: Putting Visions into Practice*, Port Melbourne: Lothian.

118

13 Cuddlepie and other surrogates*

'Home is where your fridge is': old Australian proverb. I coined it a few years back. 'Home' is clearly an artefact, whether local habitation or home country, a comfort zone we create for ourselves. All cultures strive to create a psychic homeland. How they do so varies. I am writing these words in Bali, three hours from Perth and in the same time zone, to escape the telephone for a week. Many visitors find Bali soothing. It is a beautiful and generous island, and its people are smiling and gracious. They are also animist. The rites of propitiation are sedulously, daily, elegantly served, with fresh offerings of flowers and incense at every turn. The results seem satisfactory on empirical grounds: the climate is benign, water abundant, the soils fertile, the hibiscus, bougainvillea and frangipani run riot, as much in the village compounds as in the grounds of the Oberoi. So the malign forces of the external world are kept sweet. Even the volcanic eruptions can be accommodated if you see them as renewing the soil and can spare a generation or two where they hit.

    We could try it in Australia, but I don't think it would work for us. Fire, flood and drought are our lot, and delicately woven little platters strewn with flowers are of doubtful efficacy against them. We have different rites: animist survivals are generally kept for the children, as with Snugglepot and Cuddlepie, or Blinky Bill, the cuddly koala. That living koalas piss over and scratch those who are obliged to handle them represents a much smaller gap than that between the teddy bear and his progenitor from the dark boreal forests of a now distant European past. Mr Edward had his teeth drawn long ago, first literally as dancing bear, later in his current role as bedtime familiar. Pooh Bear never had any; living only on honey, bare gums are enough.

    The garden gnome is another survival from an animist past, still sometimes seen in suburban gardens. The gnomes are an interesting phenomenon. What needs do they fulfil? The Little People of Ireland, and Snow White's Seven Dwarfs seem to be the progenitors. The Little People were taken very seriously in Ireland – beings of great power who could mediate between ourselves and the natural world, benign if

*Excerpted from an essay first published as 'Cuddlepie and other Surrogates', *Westerly*, vol. 33, no. 2, June 1988, pp. 143–55.

119

we retain their favour, malign if not. Puck, Tinker Bell, the fairies at the bottom of the garden, Pan and the dryads are all remnants of this animistic world of our ancestors, which bore many similarities to that of the Pitjantjara. But our animistic beliefs declined into folklore, became the relict superstitions of isolated rural communities, and, at length, the raw material for children's books, comic strips and Walt Disney. The garden gnomes testify that the urge to propitiate the deities of the natural world is still there, but that it can only be admitted in a joking, self-conscious way.

The Australian landscape was etched with songlines when Cook set foot on it, or Dampier and Vlaming and Volckersen a hundred years earlier, but like that of T. S. Eliot's nightingale in *The Waste Land*, the song is transmuted, 'jug-jug to dirty ears'. The rejection of animism is central to our world picture, to our modus operandi. Invoking the *genius loci* cannot be literal: the nymphs are long departed. We are left only with vestiges, embedded through the conservatism of language. The bunyip in *Blinky Bill* is an attempt to liaise two cultures at the imaginative level, but no matter how sympathetic Euro-Australians feel to the cultures that were here when Dirk Hartog and Volckersen stepped ashore, it is delusory to suppose them accessible in other than superficial and decorative ways. We try to make ourselves at home through many coping strategies. We are still learning how to cope with the physical environment of Australia: the need for good sun protection for our exposed bodies, for example, has only recently become widely accepted. Expectation based on a different environmental experience has constantly to be adjusted to actuality, of which the story of York Road poison, told in Chapter 7, Eurocentrism and Australian science, is a good example. The psychological adaptations we make are less explored. Many of the verbal ploys have already been noted in Chapter 2, Words and weeds; naming strategies, scale manipulation by zooming either up or down ('A Big Country': 'scrub')' the New Chum ploy, and the invoking of Biblical authority. May Gibbs and her bush creatures are another psychological adaptation, loved by generations of Australian children.

One of the most moving stories of accommodation to a new land and its translation into a habitable landscape is that of Georgiana Molloy, an intelligent young woman married to a man twice her age. The Molloys were isolated from 1830 in the extreme south-west of Western Australia under very demanding physical conditions, in a landscape that Georgiana found abhorrent and frightening but made home, through physical exertion, then by becoming a very knowledgeable botanist. (There are two biographies: Alexandra Hasluck, 1955; William Lines, 1994.) Her life illustrates a poem by Robert Frost about his fellow Americans:

> The land was ours before we were the land's.
> . . .
> Something we were withholding made us weak
> Until we found out that it was ourselves
> We were withholding from our land of living,
> And forthwith found salvation in surrender.
> (Frost, 1955, p. 224)

May Gibbs made her surrender to the Australian bush early, making an imaginative place for herself and for many others by creating a memorable cast of anthropomorphs, human constructs that aim to humanise the natural world, or at least link the two worlds. Cecilia May Gibbs was born in Cheam Fields, Surrey, on 17 January 1877. Her parents had both attended the Slade School and had therefore had some training in art. They emigrated to South Australia in 1881 when May was four years old. They tried farming at Franklin Harbour, and failed, so her father became a draftsman for the Lands Department in Adelaide. He next took up land, with a group which included his brother George, at Harvey in south-western Western Australia, but gave up in 1887 and went to Perth, where he first farmed at Lake Claremont. In 1889, he joined the Lands Department, again as a draftsman, where he spent the rest of his working life. After a short time in a house on Murray Street in central Perth, the family moved to 'The Dune' in South Perth, which then remained May's parents' home for the rest of their lives.

In 1900, May made the first of three return visits to England (with her mother, with whom she fought on and off), and took art lessons at Chelsea Polytechnic. She returned to Perth at the end of 1901, where she got a few assignments as an illustrator, but was unhappy and returned to London in 1904. She was back in Perth in 1905, and worked for the *Western Mail*. She escaped to Uncle George's farm at Harvey as often as possible, and this formed a major part of her early 'bush' back-ground. She returned to London yet again in 1909 – at the age of thirty-two, and again with her mother. She was employed by George Harrap & Co. for book illustrations, and met Rene Heames, who became her life-long companion. Mother disapproved of Rene. The trio returned to Australia, first to Perth, but Rene and Cecie (May's mum) clashed, so the two young women went to Sydney, and set up house at Neutral Bay. May prospered in Sydney, won contracts with Angus & Robertson and the *Sydney Mail*, and established herself as an illustrator. She lived in Sydney for the rest of her life, where she died in 1969 at the age of ninety-two, by which time she was the author and artist of eight children's books and innumerable comic strips. She kept 'Bib and Bub' going until 1967, her ninetieth year.

121

She kept in touch with South Perth, and made several return journeys by sea, including a long holiday in 1918, which culminated in 1919 with her marriage to Mr J. O. Kelly, a well-educated and charming Irishman, more than ten years her senior – May was then forty-two – virtually penniless and virtually without occupation. He became her business manager, a duty he performed indifferently, and remained a man-about-town. May was the breadwinner. They were married at the registry office in Perth during Easter in 1919, spent their honeymoon in Northam – where the bride absentmindedly signed the hotel register as 'May Gibbs' – and returned to Sydney. After six weeks in a hotel, they moved into a flat in Neutral Bay, where they were joined by Rene and Rachel, another long-standing woman friend of May.

Her character is obscure, and it scarcely emerges from her only biography (Walsh, 1985), other than that she was shy, strong-minded, unconventional and a shrewd business woman who managed to screw a 15 per cent royalty out of one publisher, the usual rate at the time being 10 per cent. However this was probably simple economic necessity – she had a household of three or four to provide for during much of her working life.

Her best-known book by far was *Snugglepot and Cuddlepie* published by Angus & Robertson in 1918. It has never since been out of print. The 'gumnut' babies had made their first unobtrusive appearance in 1913, incorporated in two separate publications. Soon after, she designed the gum-leaf bookmark, which is still on sale (I often use it as a token gift when travelling overseas):

> I thought of the Australian gumleaf, which was an ideal shape for a bookmark and a pretty thing. If only I could make it interesting on both sides. In the middle of the night I awoke, and, in fancy, saw peeping over a long gumleaf, a little bush sprite with a gumnut on its head. I hand painted them and Lucy Peacock of the Roycroft Library sold them for me at 5s each. They became so popular, later we printed them and sold thousands for 6d each. (Walsh, 1985, p. 96)

She has given the gumnut babies dual citizenship in memory: 'It's hard to tell, hard to say. I don't know if the bush babies found me or I found the little creatures. Perhaps it was memories of Western Australian wild flowers and trips to Blackheath' (p. 95). This is an interesting comment, and the evidence of the illustrations suggests that it is exact. Her drawings are not generalised but a precise record of the flora and fauna of the Hawkesbury Sandstone (Blackheath is in the Blue Mountains near Katoomba), preconditioned by an equally close familiarity with the flora of south-western Australia. These two flora are very closely related, given the distance that separates them, both very rich, and matched all down the line with the same genera and paired species. There is nothing else quite like them in Australia. The most surprising thing about May Gibbs is that she is such a fine naturalist. There is no trace of this in her background in South Perth or her trips back and forth to art school in London, and her biographer does not address the question. It was something of a Western Australian tradition for cultivated young ladies, from Georgiana Molloy to Barbara York Main and Rica Erickson, but she must also have had opportunity and some tutelage – perhaps from Uncle George at Harvey, but not, one would think, from her parents, whose culture seems to have been of the provincial salon.

Her gumnut babies seem to me to be based on the marri, *Eucalyptus calophylla*, which has an exceptionally large, woody fruit (the gum-nut), and long dark green, glossy, curving leaves with a prominent, reddish midrib, red leaf stalk, and 'fine regular parallel veins at 50–70° from the midrib. Intramarginal vein close to the margin' (Hall, Johnston and Chippendale, 1970, p. 30). These are all characteristic of the bloodwood group of eucalypts, including *E. ficifolia*, the red-flowering gum of south-western Australia, indigenous only on the south coast, but widely planted, with a nut almost as big as the marri (see colour illustration following p. 78). On the east coast, the red bloodwood (*E. gummifera*) is common on the Hawkesbury Sandstone, and the spotted gum (*E. maculata*) is also very common along coastal New South Wales. But I think her gum-nuts are based on the marri, such a distinctively beautiful tree around Harvey, the memory reinforced by *E. ficifolia* and the Sydney bloodwoods.

The Banksia Men are Western Australian, on the author's own authority:

> The Banksia Men arrived in this way. When I was out walking, over in Western Australia, with my cousins, we came to a grove of banksia trees, and sitting on almost every branch were these ugly little, wicked little men that I discovered and that's how the Banksia Men were thought of. (Walsh, 1985, p. 106)

Botanically, they fit less precisely with the banksia species common on the Swan Coastal Plain, despite her own words, although several common Sydney banksias have similar cobs, including the saw banksia, *Banksia serrata*. The wiry, curling black hairs in which the body is covered are the twisted pistils of withered flowers.

Only a few of these flowers are successfully fertilised, and these warty dry fruits or follicles make up the eyes, ears, nose and other features of the Banksia Men.

May Gibbs' comment that the image was formed in the West is illuminating on three counts. That she was out 'with my cousins' indicated that she was at Harvey, since her father's brother George was the only relative in Australia. That she had identified the Banksia Men shows that the anthropomorphic leap had been made well before she moved to Sydney; and that she saw the Banksia Men as wicked shows that she had invented that imaginative polarity of forces that drives the action of her narrative. In other words, all the primary ingredients of *Snugglepot and Cuddlepie* came from the West. Indeed, she must have become a naturalist at Harvey: she could not have done so living in inner Sydney with occasional train trips to Blackheath. (Nearly all her plants are botanically identifiable, including *Leschenaultia formosa*, and *Eucalyptus todtiana*, the pricklybark – both Western Australian; while the former is well known, the latter is not.) Her knowledge of marine biology, which is evident in *Litttle Ragged Blossom and More About Snugglepot and Cuddlepie*, published in 1920, was also probably derived largely from the Swan estuary and the Indian Ocean, although again reinforced by Sydney's coast. The *Medical Journal of Australia* reviewed this book, somewhat unexpectedly:

> Apart from her charming humour and style, Miss Gibbs is a naturalist of class. She knows every leaf and twig of the Australian bush and judging from her knowledge of sea cornets, anchovies and anemones and the like, she would seem to have spent at least half her life down in the mysterious deep.
>
> The whimsical illustrations compete for supremacy with the text.
>
> Some of them are in colour, but all of them bring a sparkle of merriment into the eye and a chuckle into the throat. We wish this delightful little volume a happy voyage in every Australian home. (in Walsh, 1985, pp. 121–2)

One other incident seems drawn directly from Perth: Snugglepot and Cuddlepie pay a visit to White City; 'They had honey sticks and dew drinks at the refreshment stall. They went on the switchback over and over again' (Gibbs, 1918, pp. 15–16). White City was a generic term for amusement parks with their canvas tents, and there was one in Sydney near Trumper Park in Paddington, but that was remote from Neutral Bay. 'White City' in Perth was at the foot of William Street, just across the water from South Perth, and by the ferry terminal.

If the genesis and perhaps the mood of her work are Western Australian, much of the detail is clearly not. Mr Kookaburra appears for instance, and koalas, neither of which were native to Western Australia, although both have been introduced. There are many distinctive plants of the Sydney region, such as the scribbly gum, *Eucalyptus haemastoma* (p. 9) which turns out to be the bush newspaper, or the New South Welsh Christmas Bell, *Blancoa canescens* (in Walsh, 1985, p. 4).

The first book is like all the later books. It is full of inventive humour – like Lilly Pilly, for example, and her father, Mr Pilly. There are some good one liners, such as the following dialogue between Snugglepot and Ragged Blossom:

> 'I'm sorry I've no clothes,' he said. 'Where I come from they don't wear any.'
> 'That doesn't matter,' said little Ragged Blossom. 'It's better to be a kind Nut with no clothes than an unkind one all dressed up.' (Gibbs, 1918, p. 20)

These are sentiments which one might expect to hear on Swanbourne beach in the 1980s, but not from a Victorian maiden lady. Also well before her day, she was a

124

The bush newspaper: the scribbly gum, _Eucalyptus haemastoma._ MAY GIBBS

conservationist, a trait which shows itself repeatedly, for example in the story and illustration for the possum caught in a trap (p. 22), but she is too good a naturalist to use vague phrases like 'reverence for life' – in fact she is very off-hand about death and killing, both of which occur with some frequency in the book. At the very beginning of the story, a greedy Owl had carried off Snugglepot, thinking he had got a tasty pink mouse. When he saw his mistake he dropped Snugglepot, who fell through the window of an 'Ant's house'.

> A tired night-nurse saw him coming, but before she could do anything he had crashed in and killed several babies. This was a blessing for Snugglepot, but it was sadly hard on the baby ants.
>
> 'I'm so sorry,' said Snugglepot.
>
> 'It can't be helped,' said the Nurse.
>
> 'What will their mother say?' asked Snugglepot, brushing tears from his eyes.
>
> 'She won't know,' said the Nurse, 'we have three hundred babies in the house.'
>
> (p. 7)

One ant more or less would hardly be noticed. The book is full of violence, but it is all narrated in an amused, matter-of-fact tone. Snugglepot and Cuddlepie sit on Mrs Fantail's eggs, go to sleep and break them all. Mr Lizard kindly brings a couple of replacements, but they turn out to hatch two young lizards. Mrs Snake is cast as a villain, along with the Banksia Men, but even she is not really villainous – snakes are merely snakes. Snakes eat small birds and frogs. The Kookaburra eats snakes.

This calm and clear-eyed acceptance of ecological reality is one of the remarkable features of the book, greatly in advance of English children's books of a similar vintage, and it has occasioned both praise and blame.

> One day I received a letter, a very angry letter, saying that it was wicked to make drawings of these Banksia Men to frighten the lives out of children. He'd torn the page out of my book where Banksia Men were on it, and scribbled it all out and dashed it with a heavy pencil, and wrote me a very nasty letter.
>
> I think getting accustomed – I mean, children getting accustomed – to ugly things like Banksia Men and that sort of thing, it strengthens them if anything, and then they find that they're not all bad, things are not so bad as they seem. (in Walsh, 1985, p. 106)

125

Most critics of children's books today would support her:

> Gibbs has a firmer grasp on the grimmer realities of life for a five year old and for us over-fives, than comparable English writers such as A. A. Milne and Beatrix Potter. The dangers and nasty characters the intrepid bushbabies face emerge from darker regions of the subconscious than Milne's 'Heffalump' or the foxes and Farmer MacGregors in Beatrix Potter. The giant squid and giant octopus, dark holes and spiders, snakes and Banksia men of *The Complete Adventures of Snugglepot and Cuddlepie* focus on strangely recognisable images; the unspoken and nameless horrors of childhood. Not the least of May Gibbs' achievements is the way, however unintentionally, she releases and copes with these images from the collective unconscious of children. (in Walsh, 1985, pp. 106–8)

The humorous control is the main feature of all these books, and they seem very sane to me. May Gibbs' own life certainly raises some questions, and there is one feature of her illustrations that is almost obsessive – the presence, on almost every page of every book, of bare babies' bottoms, nearly all males (the girls had stamenoid skirts, although these tended to ride up behind, too) (see colour illustration following p. 78). Most good children's books have come from men and women who were not cast in the common mould, and May Gibbs is one of them.

Anthropomorphs are a survival into the present of an animistic world view that seems to have value in the processes of growing up, by familiarising the unfamiliar, for children. They may have prophylactic value in giving names to name-less terrors. The most memorable children's books seem to do this – the Banksia Men belong to the same genus as the Red Queen in *Alice in Wonderland*. 'Off with her head' is also under comic control, so the delicious thrill of induced terror can be enjoyed in safety. Snugglepot and Cuddlepie themselves belong in a slightly different category, along with teddy bears: they are comforting surrogates for maternal protection, although by inversion. The child reassures teddy, and is thus reassured, and in like manner shows protective concern for poor little Cuddlepie. In short, they are coping devices, ways of coping with fear and alienation.

As for 'reverence for life' and 'conservation', these phrases are vacuous as general terms. Life consists endlessly in making choices, rarely simple or easy ones, and the art, skill and ethical achievement of a life well lived consists in making more good choices than bad ones. May Gibbs does not show 'reverence for life': but she exposes cruelty and teaches a familiarity with the bush that breeds respect.

References
Gibbs, May, (1918) 1985, *Snugglepot and Cuddlepie*, Sydney: Angus & Robertson.
Frost, Robert, 1955, 'The Gift Outright', in *Robert Frost: Selected Poems*, Harmondsworth: Penguin.
Hall, Norman, Johnston, R. D., and Chippendale, G. M., 1970, *Forest Trees of Australia*, Canberra: Australian Government Publishing Service.
Hasluck, Alexandra, 1955, *Portrait with a Background: A Life of Georgiana Molloy*, Melbourne: Oxford Universty Press.
Lines, William 1994, *An All Consuming Passion: Origins, Modernity and the Australian Life of Georgiana Molloy*, Sydney: Allen & Unwin.
Walsh, Maureen, 1985, *May Gibbs: Mother of the Gumnuts: Her Life and Work*, Sydney: Angus & Robertson.

14 Jet-set and parish pump*

The global village and the rebirth of regionalism

In the last few years I have watched two apparently opposite trends, and although one is partly a response to the other, I do not understand the relations between them. One is the move towards the so-called global village (McLuhan and Powers, 1969), a world made small by good communications and homogenised by technology, so that citizens of Kansas City and Melbourne and Marseilles shop for the same goods from the same supermarkets and watch the same TV programs. The other is a rebirth of regionalism, which emphasises the differences between communities rather than their sameness. Both are very striking social phenomena.

As manifestations of the rebirth of regionalism, consider the following; first the legitimisation of regional accents on the BBC after World War II. Before about 1950, almost all the BBC announcers had 'BBC' accents, that is, an educated London accent, whether they were broadcasting in London or Yorkshire. Now, regional accents are *preferred* for the regional networks. Before the war, an Australian accent was never heard on the Australian stage or any other, unless in vaudeville, and it was a marked disability at the higher professional and social levels of Australian life. But the right to speak with one's natural accent, so recently won, is part of a more general move towards linguistic regionalism. After nearly two thousand years of Latin, the Catholic Church turned to vernacular liturgies. Scottish and Welsh nationalism are again rampant in the United Kingdom (United? Kingdom?), and this has a large linguistic component. These movements are not nationalist in the grand nineteenth-century fashion, the fashion that welded Italy and Germany into nation-states. The Scottish and Welsh movements are regional movements, and such movements are ubiquitous – there is even muttering from the Cornish. French-Canadian regionalism has reasserted itself after nearly 150 years of near peace. The Flemish and the Walloons are split in Belgium (Chechnya and

*First published as 'Jet-set and parish pump', *Quadrant*, vol. 17, no. 3, 1973, pp. 46–52.

Kurdistan are more recent examples); periodically renewed agitation for secession in Western Australia – not, perhaps to be taken seriously, but last heard of during the Depression.

That the global village is closing in on us hardly requires exemplification, although perhaps the metaphor needs scrutiny, as does its companion metaphor, 'Spaceship Earth'. The earth is not a spaceship, and it is certainly not a village. It is unlike a spaceship in many ways: for instance, we did not make it, we do not control it, other than in minor ways, and we do not understand many of its major operations. It is a complex and interesting place. The one thing Stanley Kubrik's film *2001: A Space Odyssey* brought home to me is that space travel in the future, like jet travel today, is going to be very boring, whereas our planet is anything but boring. Our significant journeys are not on it, through space, but around it, our own world. Nevertheless, the metaphor is a powerful one, if glib; it reminds us that apart from the input of solar energy our planet is a closed system – it is *our* atmosphere, *our* hydrosphere, *our* biosphere; they are the only ones available to us, hence it behoves us to look after them. Concerns such as these force us to expand our loyalties to embrace the planet as a whole ('Vote for your friendly neighbourhood planet'), and seem therefore to be opposite in direction to the centripetal forces that contract our loyalties to the parish pump.

In what sense do we live in a global village? The world is certainly a much smaller place than it used to be, at least for the affluent who fly hither and yon; and, I suppose, for a few trendy professors, the round earth's imagined corners may seem none of them out-of-the-way, and all as cosy as Greenwich Village, Georgetown or Chelsea. I don't suppose this experience to be universal, in that a New Guinea man or woman from the Sepik River would probably not find Kennedy Airport especially cosy, and they would not consider it to be part of their village. But then their problem – the one we have set for them – is to make the national scene, a phase that Western societies passed through a hundred years or more ago.

In one of the *Dr Who* series on television, armies from different periods of history are brought together by a time machine, with the result that Roman legions join battle with Hitler's storm troops, Scottish Highlanders with Napoleon's army, and tribes of cavemen with the Grenadier Guards. One effect of tel-star television and internet-fed newspapers is like that evil plot. Peoples around the world are living, as they always have, at very different stages of technological development. The effect of 20th century communications is not only to juxtapose far places, but also, in effect, to throw different epochs into stark contemporaneity. As a consequence, students of the London School of Economics are fighting the battles of Brazilian peasants, while Asian peasants are thought to be fighting for the ideas of German romantics like Marx, and Bantu tribes in Africa are represented as demanding John Stuart Mill's reforms. The pseudo-contemporaneity imposed for the viewer by television on culturally remote societies leads us to forget our intellectual limitations. Few of us understand our own culture and its aspirations, let alone those of Vietnam or the Bantu.

Thus I do not think talk of a global village is appropriate. The world has not grown cosier for most of us, although we do know more about it, at least in a superficial sort of way. Radio, television and the newspapers keep us informed about politics in Chile and drought in Ethiopia. The news is not all trivial, and some of it increases our sense of interdependence. Market fluctuation on Wall Street and in Tokyo make this very clear, and it takes no special prescience to see that the health of the Japanese economy is reflected in the price of Australian wool and minerals, although not necessarily directly, because contracts may be written in US dollars. The force of this again is to heighten our sense of the earth as a closed system, because what happens to one country has an impact on all the others. This common recognition of interdependence is new, I think. In the nineteenth century there were the big, powerful countries who did things, and the small or powerless countries who got things done to them. Whether or not it was really like that, it seemed like that, but it doesn't seem like that any more. The powerful countries are so vulnerable. For instance, it seemed likely (in 1973) that in the last two decades of the twentieth century, the price of Middle Eastern oil would be a critical factor in the future of the technologically advanced countries, especially the USA (and the Gulf War has borne this out). Such considerations act centrifugally on our consciousness, in the sense that our concerns are expanded outwards to embrace the globe as the significant unit. But a predicament arises. Our intellectual recognition of and emotional commitment to the proposition that there is only one earth and that we are all on it together are powerful, but they are not given much outlet in effective action, because the machinery for achieving the kind of far-reaching co-operation we now need is lacking or rudimentary, and I have no doubt that this is part of what sets people to cultivating their own gardens, as they have always done. The retreat behind the monastery walls in the Middle Ages and the hippie farm communes in North America and Nimbin in northern New South Wales have some obvious points in common. But I doubt that this is the whole story.

The effects of technology

Another part of it is clearly the homogenising influence of technological society. Alvin Toffler (1970) in *Future Shock* tells us that mass production, with its dreary uniformity, was a feature of early industrial society, but that we now have got to a new phase, post-industrial society, which is characterised by an enormous plethora of choice. This thesis is not entirely vacuous, but much of the choice is illusory – there may be thirty different kinds of breakfast cereal on the supermarket shelves, but they nearly all taste like shredded brown paper, while there are only about four or five kinds of apples, where there used to be Cox's Orange Pippins and Snowflakes as well as Jonathans and Delicious. An Australian can choose between forty-odd makes of car, but he can't buy one that is safe, or non-polluting. It is at least difficult to get one that is durable, reliable, and reasonably priced, and it is little consolation that there is a wide range of choice in the fabric and colour of the upholstery. The homogenising influence restricts choice, and substitutes for it, pseudo-choice. Moreover, it does not operate only on manufactured objects, because through them it largely determines how people live. This is most apparent at airport terminals, which are the same the world over, but their influence extends to the world around them, and it can almost be said that if the point of travel is that your destination should be different from your starting point, then it is no longer possible to travel by jet – in fact I sometimes feel that the only genuine tourist experience that the world has to offer is travel in the USA, because that is the only place left where the natives are not putting on a show for the Yanks.

129

The effect of all this is to reduce the richness and diversity of the world and its cultures, to make us more alike, smooth out national and regional idiosyncrasies. Marx foresaw and welcomed all this as one of the great achievements of the bourgeoisie. Industrial production in the nineteenth century had already 'rescued a considerable part of the population from the idiocy of rural life' and would in time spread the same type of consciousness throughout the world. The peasants' mode of production differed from place to place, and so their consciousness differed from place to place, too. A friend of mine who is a political philosopher was driving through Germany on an autobahn recently when he saw a rest-place called Wupperthal Gasthaus. In his own words,

> Now Wupperthal is Engels' birthplace, so I stopped to drink his health. I do not like Coca-Cola, but I thought this would be the only appropriate drink to his health. It is produced in exactly the same way from Japan to Germany, from Cairo to Sydney.

A good wine, a Moselle, for example, would express only the partial consciousness of the peasants of the Moselle valley or a different wine, the partial consciousness of another valley. But Coca-Cola is the material equivalent of the Universal Spirit, it is the Incarnation of the Objective Universal Consciousness. Nor did he get it at the counter, because Engels had said that one should not employ a man to serve another man; he got it from one of those automatic machines in front of which we are all Equal. 'Was that the unalienated moment of my life, the experience of the Global Village that I should cherish, as mystics cherish their rare unions with the Absolute?', asked my philosopher friend.

The drive towards the One True Consciousness is manifested in many ways in our society, all the way from Marxism to unisex. There is no such human aim or activity as being just 'Man' – this is one aim of the mystics. If it were to succeed, then it would be enough if one of us existed. (If you have slightly different thoughts and values from me, then, *ex hypothesi*, we have not yet obtained the one true consciousness.) It does not even matter which one; one is enough and two would be indistinguishable. Medieval thinkers were familiar with the problem, in a slightly different guise: what distinguishes one angel from another in the Beatific Vision? (They found the only possible solution, to argue that each angel is a different species, so that the difference between one angel and another is like the difference between an aardvark and an armadillo, and not like the difference between two aardvarks.) So in the Heavenly global village there need be only One Unisex Being.

Evaluating regional culture

He/she won't have much to do. It seems obvious to me that one can only achieve one's aims and intentions if they are shared with others and there are institutional opportunities for achieving them; and that a complex society offers more opportunities than a simple one. To that extent, I understand Marx's contempt for 'the idiocies of rural life'. But he despised the complexities of regional life, and exalted the idiocies of industrial homogeneity. Fortunately, moves in the other direction are just as clear. Coca-Cola sells well, but so do the wines of the Moselle, and the regional character of good wines is perhaps more highly valued than ever. Once again, this is partly reaction. When our identity is threatened, because we all become more alike, we cling to rags of difference, no matter how tattered. But this depreciates what I am talking about, and it might be put differently. Consider the impact of our first trip outside our own country, for example Australians arriving in London. We become conscious of being Australian in a way that we were not before. How this new consciousness finds expression depends on the personality of the Australians: they may cling together with other Australians for reassurance; they may subscribe to the ready-made myths of the Australian versus the Pom, for example, 'Aren't the Poms dirty buggers, they only take a bath once a week', etc., and call for ice-cold Fosters; they may feel ashamed of their superficial Australianisms, and do their best to pass for natives. They may do none of these things, but still be stimulated in all sorts of ways, for instance by the very different light intensities of England and most of Australia, and the effect this has on perception; or perhaps by the feeling that English soil has been trodden by so many feet for so long, and relish by contrast the relative 'freedom from history' that the Australian is sometimes said to enjoy. My main point is obvious, that travel provides contrast, contrast heightens perception, and this has both crude and subtle manifestations.

But if this holds for Sydney–London, how about Sydney–Melbourne? Do Sydney and Melbourne become more alike as communications and travel opportunities increase and as technology gives us the same cars, in fact an almost identical range of consumer goods? Are they becoming submerged in the global village? It is hard to talk sensibly on a topic like this, because one immediately gets caught up in one of Australia's most popular parlour games, although to outsiders it might seem the Melbourne–Sydney rivalry is like that even more intense rivalry between Oxford and Cambridge, which are culturally interchangeable to all but the participants. There is surely a remarkable degree of homogeneity between Australia's large cities – despite their geographic spread, they are far more alike than are Boston, New York, Chicago, Philadelphia, New Orleans, San Francisco and Los Angeles. Is the Melbourne–Sydney rivalry then anything more than a parlour game, or to put the question differently, does the rivalry spring from a sense of difference, or a sense of likeness? And is it waxing, or waning?

There are, of course, real differences between the two cities, most notably in topography and climate. They differ in their social origins and were born in different centuries. They play different games (or did). Fort Street and Sydney High have educated Sydney's elite; Melbourne Grammar, Melbourne's. Melbourne is (or was) the financial capital, but Sydney has a prettier zoo. To go on would be to play the parlour game. The question is, are the similarities, so obvious to an outsider, superficial, and the differences profound, or vice versa? Are we inflating differences because we know in our bones how alike we are all becoming, or has ease of communication heightened our perception of significant cultural difference? I don't know; but on the whole, I think the latter. I think people choose to live in Melbourne or Sydney – those who are free to choose, of course – at a high level of consciousness, and their choice is in accord with their temperament, and thus I suspect that mobility accentuates difference, in that Melbourne-type people gravitate to Melbourne, while Sydney-type people levitate to Sydney. People are loyal to the idea of their city.

Loyalties and commitments

Loyalty is part of what we are talking about. Societies change through time in all kinds of ways, and one of the most interesting is the way in which people distribute their loyalties and allegiances. What evokes loyalty today? Clearly, this depends on the people and the place, but perhaps some generalisations are possible. For instance, there is not much loyalty in Australia today to the flag or the idea of the flag as a potent symbol. State Schools in New South Wales have been instructed that they are to fly the Australian flag. It is fairly clear from the response that to most people in this state, and especially the young, it is a matter of indifference whether or not the flag is flown, with a faint preference for not. To some old-timers, it may seem that the young are no longer loyal to the flag, and that this indicates a general breakdown in loyalty ('They don't believe in anything nowadays'), whereas what has happened is that loyalties have been redistributed. We usually withhold loyalty when we suspect that it is being incited by administrators for their own convenience. At my son's primary school, the children were divided into four Houses or factions with names like Banksia and Wattle. This is a faint Australian echo of the English public school in which the boys board in different Houses, originally the private house of the housemaster. It was maintained in Sydney day schools as a handy way of dividing the children for sport, and as a way of manipulating behaviour, since bad behaviour loses points for your faction. Many of the children are quite indifferent to this '*divorce de convenance*', but some of them are taken in by it, and several teachers think that pupils who are not loyal to their faction show a deplorable lack of team spirit. Attempts to manipulate our loyalties when we are young probably tend to make us wary of institutional loyalties in later life.

The degree of loyalty that is accorded to institutions will obviously vary with the place, the institution, and the person. The Japanese seem to be successful in inducing loyalty to an employer, but this is because their big business houses are extraordinarily patriarchal, caring for the employee and his family from the cradle to the grave, and perhaps the loyalty is in some degree compelled rather than a spontaneous commitment. American firms try hard, and their public relations experts promote the image of, for example, Mother Bell, the kindly old lady of the Bell Telephone Co., but nobody takes this seriously. Australian firms until very recently didn't even try – it never occurred to them that their employees could or even should be loyal to them, and they weren't smart enough to see that it was to their financial advantage, although they know now.

What about loyalty to schools and universities? This again must be a very personal matter, and it must depend on the school or the university, but I suspect that this sort of institutional loyalty is at a fairly low ebb. Although I was intensely loyal to my old school and to my undergraduate university, I must admit – perhaps unwisely, but honestly – that I had almost no *feelings* at all towards the University of New South Wales (where I was a professor at the time of writing) as an institution, nor, I think, did most of the people around me, except perhaps for the few who had seen it grow from the egg. We did, of course, get pep talks from the senior administrators from time to time but they seemed to most of us to be on the Mother Bell level. The operative loyalties that I discern around me are either very local, to the school, or international, to the profession.

This professional loyalty, of the historian to historians, and geologist to geologists, wherever they may be, are examples of a more general trend in which the sense of community goes to the peer group rather than the institution, and this kind of commitment is very striking with the young. For a whole generation, the Beatles commanded intense loyalty, one that was far stronger than ties to school or country. The effect of good communications has for some been standardising, but for others, liberating. Through the telephone, rapid transport and good mail systems, we can, if we choose, be members of many different societies that may be physically dispersed and yet have the warmth and intimacy popularly associated with village life. Each of these groups may give social expression to different parts of our character; if we are restricted by circumstances to one or a very few groups, then large areas of our character will never find social expression, yet the very diversity of contact available to us today is suspect to those who yearn for the One True Consciousness. If we are so many things to so many people, where is the Whole Person, our simple self, to find expression? Nowhere, of course – but the Whole Person is one of those romantic dreams, like the Doctrine of Original Virtue, that has made so many people so needlessly unhappy. Anyone who thinks that the 'Whole Person' gets expressed in village life knows nothing about village life, where the first rule is to remember who you are talking to.

Neighbours and neighbourhoods

We belong to many different groups for different purposes, and some of my neighbours live thousands of kilometres away. Thanks to technology, there can now be small, select villages connected not by the town crier but by air-letter (and fax, electronic mail, computer networks, the Internet); their membership is voluntary, and it does not preclude membership of other villages differently constituted. There are innumerably many different little groups. One of these, but only one, is the group of those who share no ideas and values other than the idea of a global village, and they do not communicate anything other than what is communicable on television. They live in a village in the worst sense of 'village', in that they are so provincial that they do not know that there are people besides themselves, sharing many other values by many other means. It is because such patterns of shared interest so often and so easily cross national boundaries that such boundaries seem arbitrary and fortuitous rather than vital forces in our emotional lives. Nationalism surely is in decline in general, although it is surprising how easily it can be revived when it has some urgent content.

133

But there is no decline in small-group loyalty – I spoke earlier of a revival of Welsh and Scottish regionalism. For my last example I would like to take something nearer to home, the rise of the resident action groups. These are partly political, in that a group of people band together to defend what they regard as their common interest. But they could not do that unless there were some perceived limits to the area concerned, and a set of agreed common interests. I belonged (in 1973) to the oldest, and one of the most powerful, of these groups in Australia, the Paddington Society. What is it about? Paddington and some of the other inner-city suburbs have attracted a few bohemians for many years, and the Paddington Society began life as an informal social group. It came to life politically when it successfully fought an attempt to widen Jersey Road ('to widen Jersey Road would be to destroy Paddington').

In the 1948 County of Cumberland Planning Scheme, Paddington was designated as 'totally substandard, requiring replacement', and the intention to demolish and rebuild was still very clear in the City of Sydney Planning Scheme of

1958, by which time redevelopment had begun. It consisted in pulling down rows of rather cramped and near derelict terrace houses and replacing them with very much more cramped blocks of flats, with no increase in street widths, or green space, or playgrounds, or school facilities. A cosy, human old slum was being replaced by a cruel new one. But by 1968, this slum was scheduled as 'a special area of architectural and historic interest'; flat-building and demolitions had ceased, and many of the terrace houses had had thousands of dollars spent on their restoration. Houses that sold for £3,500 in 1963 sold for $35,000 in 1973, and the better terraces for $50,000 to $80,000. (In 1996 they would fetch at least $500,000.) An estimated $2 million, all out of private pockets, was spent on restoration between 1959 and 1965, and more than $5 million in the next eight years. Why? Mere fashion, in part. But people must feel they are getting something for their money, and what they feel they are getting is urban amenity. The migrants perceived this before the trendy upper middle-class Australians, especially the Italians, Greeks and Portuguese, who bought their terraces in the 1950s, and put them in order. Terrazzo sprouted, as did wall plaques with a Mexican nodding under a palm tree by the front door. But they also fixed the roof and swept the street, and this was the major step in the recovery from dereliction. Next, the architects and interior designers, university professors and successful writers moved in. As they and others like them had sunk quite large amounts of money into their terraces, the Paddington Society became, in large part, an association of the well-to-do to protect their real estate investment. The Italians could not resist the high prices, and began to move out. But a cynical account of this upper middle-class takeover cannot conceal the fact that people came – and stayed – because they loved the place.

What they liked, mostly, is its individuality. The area is a sharply defined topographic unit, an amphitheatre bounded by the ridges of Oxford Street, Darlinghurst, and Edgecliff, facing north, sunny and sheltered, and visually all of a piece. It is compact – you can get round it on foot. It offers great human variety. But its most striking feature is that it is about as close-knit a community as any rootless, mobile, transient urban community of this century can get to be. There is so little private open space that a lot of living gets done in public, as in most European cities, but this is in contrast to the Australian suburban ideal of inviolate privacy behind the privet hedge or fibro-asbestos back and side fence.

It may be that the upper middle-class takeover in Paddington will in the end be so complete that the human diversity will decrease, and that it will become a sort of south St Ives on small lots. But I don't on the whole think this will happen. I think that St Ives-type people will become more so, and that Paddington-type people will become more so, and that this is on the whole a good thing, because people are different, and because Sydney is quite unmanageable and incomprehensible as a whole, except for purposes of berating Melbourne.

So, to conclude, I think Mr Toffler is quite likely right, that there will be more diversity rather than less, but I don't think this entirely for his reasons. As for Mr McLuhan, I seem to have cast him as the global village Idiot. We shall have to think 'Whole Earth' thoughts in the years to come, and somehow translate them into 'Whole Earth' deeds. But there won't be any global village in the immediate future. Perhaps we are living on cultural capital: one could put the case that St Ives and Paddington have become more sharply differentiated because the Italians and

Greeks moved in to Paddington, and the well-to-do Australians who followed them were already familiar with European cities, and knew that there were alternatives to the suburban dream – and so the apparent increase in diversity is merely a late colonial reflection of the historic diversity of European cultures. This may be true. We may be living on capital. At least I am sure that we can do so for some years to come.

I referred earlier to Kubrik's film *2001* and space travel. One sharply observed moment in that film showed three astronauts sitting in the lounge of the Space Hilton, waiting to go off on their various missions. One was a French girl, one an Italian girl, one an American boy. The moment consisted in a quick private smile exchanged between the two European girls at the bouncing puppy-dog enthusiasm of the clean-cut young American male. I have no doubt that that smile will still be smiled in the year 2001, if any smiling lasts so long, although perhaps by then it will seem anachronistic to speak of an Italian girl, because we will see her as a Florentine or Venetian or Roman. Perhaps this is the right image for the future, a Florentine in a space-suit, an inhabitant of the global village, yet maintaining a strong and healthy regional culture; as in the Middle Ages, when loyalties were given to Christendom on the one hand, and to city-state or principality on the other.

References
McLuhan, Marshall, and Powers, Bruce, 1969, *The Global Village: Transformations in World Life and Media in the Twenty-first Century*, New York: Oxford University Press.
Toffler, Alvin, 1970, *Future Shock*, Oxford: Bodley Head.

15 Placing the debate: a long postscript

I have added a long postscript to these essays, partly because three of them are early (1972–73), and partly because ambivalence runs through them, unresolved; but the ambivalence runs deep in our culture. The opposing trends are explored in 'Jet-set and parish pump'. There is little difficulty if 'sense of place' is construed solely in biophysical terms, as it is in Chapters 11 and 12. In my book, *Sense of Place*, the last section is called 'Man', but much of it is urging the people of Perth to pay more respect to their natural environment. When 'sense of place' is used to urge the recognition of the cultural identity of a town or region, as is quite often the case in articles on urban design today, the going gets harder. I concluded my book with the words: 'There is nothing wrong with being parochial if your parish is a good one', but this was both a part of the 'small is beautiful' rhetoric of the day, and a hope that Western Australians would come to respect their physical environment rather than continue the attempt to turn it into something to which it is not well adapted – and to persuade them to be less defensive. It was not meant to sanctify the mindless local patriotism that is the obverse of a deep-rooted sense of inferiority.

The cultural component of 'sense of place' raises four problems for me, some of them touched on already. The first is that it must raise the question of 'whose place?'. There is not much discussion of Aboriginal Australia and none of multicultural Australia in Chapters 11 and 12. The second is the awareness that in the age of mass tourism, the regionally distinctive has a cash value. Examples abound, and so do their costs. All houses on the Isle of Anglesey in North Wales have to be painted white. The outcome in terms of environmental design is admirable, the motive less so. It is what the tourists like, and tourism plays a big part in Anglesey. I remember the heartfelt cry of a professor at a conference in Vienna that 'We Austrians will soon be compelled by law all to wear lederhosen and take lessons in yodelling'. This leads to the more general problem: how should proposals to express or reinforce the 'sense of place' be evaluated? This is a key question for the design professions. Is the Big Banana at Coffs Harbour, or the Giant Pineapple that greets the visitor at Nambour, a legitimate expression of communal economic identity or a blot on the landscape, or both?

These questions about the evaluative application of the concept are explored shortly, but first I want to expose my own ambivalence, and to give it an auto-biographical setting. First, I am torn by two cultures. One is the academic culture, which is international, the enemy of the parochial; its habitual mode is ironic detachment. But, like many of my colleagues, I am also prone to bouts of passionate commitment; indeed, as I said earlier, I have long been 'the sense of place' man, called on to speak at conferences when that concept is on the agenda. Why have I spent so much of my life wrestling with this concept? I can think of three answers, all of them probably true. The first is favourable to me: that I have done so from very real conservation concerns. The second is more probing, but a fair question. Is this a psychological need that follows from the extraordinarily peripatetic life I have led, both in Australia and beyond? Is it compensatory, especially for a vagrant childhood, never in one place for long? If this is true, as it may well be, then it is also true of an unsettled culture, almost world-wide. The third, contrary, answer is that travel can *strengthen* the sense of place. 'What can he know of England who only England knows?' One can make a distinction between the local or regional commitment that springs only from limitation, knowing no other, and the kind of commitment that follows conscious choice, knowing a wide range of alternatives. I have discussed the probability of personal predilection in the introduction to Theme III; here is the personal journey that led up to the writing of *Sense of Place* in 1972 (looking back from 1996).

My own journeying

My father was a country bank manager in Victoria, and we were moved around. I was born in the Mallee, moved to Romsey, then Heathcote, then Mildura, where I went through most of my primary schooling. I also began high school at Mildura High, until I won a scholarship to Ballarat Grammar, where I finished my school years; meanwhile, 'home' moved from Mildura to the Wimmera, first Nhill, then Horsham. Mildura was unlike the other towns, physically, socially and historically part of a different Australia. Heathcote was a gold town of the Central Goldfields of Victoria, and the others also were a response to the gold years, growing food for the new arrivals. But Mildura was not mid-nineteenth century, but turn of the century and later. Most of the buildings were fairly new, including our bank, which even had glass bricks. The settlement was a green chequerboard laid out on the dry red sands of a semi-arid pastoral landscape, reckoned in acres to the sheep rather than sheep to the acre. The Murray was its life-line, and we spent as much time by it and in it as we could, swimming bare down by the lock, loving the muddy banks, grey water and the river red gums, one of Australia's great trees: I have seen them since by the dry sandy bed of the Gascoyne and Murchison, and along Coopers Creek in Central Australia, huge trees full of corellas; and used in a regeneration program on the mined bauxite sites at Weipa in Cape York – everywhere except in south-western Australia, where *E. camaldulensis* is replaced by *E. rudis*, the closely related flooded gum, starting around Greenough.

Perhaps my nomadic academic life is a product of these early days, although my mother says that I was always restless. It has forced me to be adaptable and to avoid that way of being partisan which commits to one place at the expense of all others. I find the surface of the globe endlessly fascinating, and no longer find any part wholly dull, although of course I like some places more than others. My father once told a story of a business man from Gippsland who was thinking of moving to the Wimmera. 'Are the people here easy to get on with?', he asked my father. 'Do you find them easy to get on with in Bairnsdale?', said my father. 'Oh, yes', said the man, 'they are a very co-operative lot'. 'You will find them very friendly in Horsham, too', said my father. Places are easy to get on with if you meet them on their own terms.

But what are those terms, and how does one discover them? That is what these essays are about, and of course there is no answer to such glib questions, so the essays are an exploration. Yet I have learned something over the years. Western Australia taught me the importance of the scale of attention, since so much of interest here is the fine detail: the incredible delicacy of the leaf form in most of the Proteaceae, for example, especially with *Dryandra* and a banksia like *B. brownii*; or the exquisite, tiny rosettes of *Drosera*. That balance of foreground, middle ground and background with composed masses to the right and to the left that make up our inherited sense of the picturesque hardly work in banksia woodland or jarrah forest or indeed in much of Australia. That is one reason why 'native' gardens are hard to design, the parts usually being more interesting than the whole.

I came to Western Australia from the University of Toronto in 1956, and my first responses are told in Chapter 11: this was not at all what I had come 'home' for. During the next decade, I became a fairly loyal Western Australian, ambivalent, of course, and an acidulous critic, like most of my university colleagues, few of whom were born in Western Australia or even in Australia, and many of whom were barely committed to either. But I began to grow a fierce love of this dry land: when I flew in from the east after a conference, I used to look out at the salt lakes below, and the desolate, scabby wheatbelt, and say to myself: 'This is my country, these are my people', electing to despise the gentle, easier landscapes of the 'fertile crescent' along our eastern shores.

I slowly began to be an Australian, rather than a Victorian. I worked for Western Australian Petroleum for a time in the West Kimberley, which was a great privilege, since it was otherwise inaccessible in those days. The hotel at Fitzroy Crossing had a sign that read 'Gentlemen will please wear a singlet in the dining room'. We also drove the Nullabor quite often, and so began to get a feel for the continent.

Other learning came from travel outside Australia, of which I have been able to do a great deal, although mostly after I left Western Australia or since my return to it in 1988. I have learned, and am still learning, about design from Italy, while my understanding of the extent to which landscape is a cultural product has come most acutely from travel in China and South Africa.

China helped me to learn about focus: Chinese books and landscape paintings are in scrolls, unrolled to be read, in linear mode. I have one of the Yangtse Kiang that follows the river from source to mouth. The focus of our own inherited culture is learned from the Golden Mean of the Greeks, that sweetly proportioned rectangle to which most of our paintings conform, if only roughly. Our sense of its inevitability is massively reinforced by the camera: we compose through our view-finder, looking for informal balance within the prescribed frame. Fred Williams broke the frame and painted in strips, like Flinders before him. Australian land-scapes are seamless. They rarely compose so neatly into identifiable 'scenes'.

Africa, the most likely birthplace of our species, helped to make me aware how long and pervasive is the interaction between people and landscape; there is probably no land, even Antarctica, that is not modified in some way, sometimes for the worse, but also sometimes for the better. It is now very clear that the Europeans who entered this continent came to a humanised landscape. If it suited their cattle and sheep, it was because of centuries of grooming for herbivores. The land was neither empty nor primeval.

Placing 'Sense of Place'

When my academic voice takes over from my personal one, I reflect that 'sense of place' is, in general terms, a part of the vocabulary of romanticism, in so far as the classicists value generality, and the romantics, specificity. The preference for one or the other in a given culture depends on the mood of that culture at a particular time. The mood can usually be assessed by the antonyms in use. If 'sense of place' is valued, then its opposite will be referred to as 'placelessness'. (There is an excellent short book by E. Relph, 1976, called *Place and Placelessness*.) People who lack an identification with place are said to be 'rootless'. Obviously this is a metaphor, although those who use it often forget this. We are not trees, and we do not need roots, either for sustenance or stability. We are a highly mobile species, and always have been, with a past of hunting and food-gathering over quite an extensive territory. Some people and some societies, however, undoubtedly do draw sustenance from their identification with place; the question thus becomes 'what?' and 'how?' In 1991 I published 'Articulating a Sense of Place', and the following paragraphs come from that essay.

A confident metropolitan society is unlikely to value 'sense of place', and will oppose the word 'parochial' in contrast with their own metropolitan culture. Such confidence has been in retreat generally over the last decade or so, even in New York. The revival of romantic values has always been closely tied to perceived urban-metropolitan crises. Wordsworth fled London, but he was not the first. That celebration of the minutiae of the natural world that is such a strong tradition in Chinese culture is the work of urban civil servants writing and painting in their country retreats, and Vergil wrote his Eclogues and Georgics far from the Rome from which he drew his income, detailing the life on his farm in the Euganean Hills. The 'pastorale' is an expression of urban culture, and the countryside so celebrated has always been seen from the security of a hobby-farm, from which the writers do not have to draw full sustenance, and to which they are not chained.

In our own times, the concern with the homogenising effects of international technology and production has led to a renewed concern for the locally distinctive. The question that needs addressing is that of articulating adequate criteria for distinguishing between valid and specious expressions of that concern. Consider the Norfolk pines of Cottesloe, where the Council is now planting many more. This recognises that the pines are part of the 'identity' of Cottesloe, and that they express a 'sense of place'. But when they were first planted, they must have been totally out of character with the landscape of low, stunted scrub wholly without tall trees. Thus this concept is a very plastic one. Since the Norfolk pines would have seemed quite alien to the Aboriginal inhabitants, we must again ask *who* is identifying with the place.

139

Moreton Bay figs breathing the very spirit of place at Rottnest. They are as much displaced from their natural environment as were the Aborigines from the Pilbara and the Kimberley who were imprisoned there, roughly at the time that the figs were planted. MICHAL LEWI

140

Many useful concepts are nevertheless uncertain in their application. In this case there are two levels of uncertainty. The first is noted above; more fully, we can say that design, or literature, or art that evinces a 'sense of place' respects local context. But, as we have seen, 'context' can be interpreted both physically and socially, and the two interpretations may be in conflict. An interesting example of social conflict caught my attention some years ago in the Snowy Mountains. Between Guthega Dam on the Upper Snowy, and Island Bend downstream, there were several construction camps for the workers, and in a small campsite, long since abandoned, I found a glade of lupins growing wild under the snow gums. It was a charming sight, and not without pathos. Many of the workers on the Snowy Mountains Scheme were, as I mentioned in Chapter 5, what we insensitively called 'DPs' (displaced persons) often from the Baltic States (or 'Balts'). Some of them must have planted a few Russell lupins, the garden lupins, around the entry to one of the stark prefabricated huts, as a reminder of their distant land. There could be few other places in Australia with a mild enough summer for garden lupins to 'naturalise'; here they had done so, in a small cheerful clump of soft pastel colours. The policy of National Parks and Wildlife was to rid the mountains so far as possible of introduced plants, to protect the integrity of the natural environment. However, their charter is, broadly, to conserve both our natural and our social heritage. Here the two were in conflict, but the natural scientists were in the ascendant, and to them, the lupins were 'out of place'. Once again, '*whose* place?'.

But the second level of uncertainty is more basic. Why *should* we design in context? In architecture, the Bauhaus and the International style produced some very good buildings that proudly proclaimed their non-local character, and some of their best and most dramatic effects are achieved by contrast, for example, the I. M. Pei towers on the edge of eighteenth-century Philadelphia. Unrelieved harmony can cloy: a little discord can heighten our sense of harmony, just as the international can dramatise the local.

Designing in context can very easily yield a bland mediocrity. It appeals to many planner-administrators because it makes relatively simple rules. In rural areas, for example, especially if they are near a large city and attract the Pitt Street farmers (or Collins Street, or whatever), one now often encounters a requirement that new buildings should have roofs of low pitch and of non-reflective surfaces, and the walls should be of natural material or painted in earth colours or olive green or the like. The objective is that they be unobtrusive, appropriate to their rural context – yet the best building in one such area, the Mornington Peninsula, is the Cape Schanck lighthouse, a strong vertical accent, stark white and intensely visible, which is, of course, its function.

Most Greek villages, churches and all, are whitewashed, too, and perfectly in context in their own setting. Of course the regulations are devised to prohibit the grotesquely inappropriate farmhouses that have sprung up so readily, red-brick triple-fronted bungalows, indistinguishable from the standard of suburbia, but we cannot, alas, create good taste by legislation. Can you imagine the Parthenon getting planning approval today in one of those rural shires? The site is too prominent, to start with. The temple would have to be built below the top of the hill, to avoid 'skylining', and painted in tasteful shades of pale ochre or olive, and screened with a dense protection of native vegetation. Planning approval would doubtless also require parking for 200 chariots, and the use of marble would probably be prohibited, at least from the open quarries on the hillslope visible from the city of Athens to this very day, great scars on the landscape.

The timidity reflected in such design and planning guidelines is typical of our day. We have lost confidence in our builders and architects: we fear, with good reason, that if left to themselves, the results will often be horrendous, so we try to regulate them, the paradox being that this is also inimical to good design. The nostalgia cult is part of the same phenomenon; the past appears to have a great future as colonial homes spring up in all the new suburbs. Since many genuine colonial buildings had a simple elegance, well adapted to our climate, this is perhaps not a bad model, provided that sound function and good proportions are of the essence (see colour illustration following p. 78). But in practice, the focus is usually on the superficial detail, so characteristic of the fashion industry. That may not seem to be a satisfactory conclusion. There is none. One can suggest many things that a good designer should consider, but one cannot prescribe good design.

And so the debate goes round. Although it cannot provide a clinching argument, the concept 'sense of place' can at least help to stimulate reflection about the nature of both the cultural and physical context, necessary in fully understanding and evaluating art of any kind. I shall complete this sketch of the concept-in-action with a quick look at the glass pyramid designed by I. M. Pei as an entry to the Louvre in Paris. This building at first outraged many Parisians, but most of them seem now to be proud of it, as they should be, because it is a fine building. The shock of the new caused the outrage: had I. M. Pei no respect for the setting among the neo-classical buildings that constitute the Louvre as we had known it? Had he no sense of place? But Paris was once a Gothic city: the first buildings of the Renaissance must have seemed equally incongruous when they were built, and the Haussman boulevards, even the splendid Champs-Elysées, which were sliced through old Paris, must have at first seemed a savage butchery, although they now breathe the very spirit of Paris. One can argue – and I *would* – that I. M. Pei has understood Paris well: he has given it the uncompromisingly new, but done so with flair and élan, thus expressing the best tradition of this great and always innovative city.

References
Relph, E., 1976, *Place and Placelessness*, London: Pion.
Seddon, G., 1991, 'Articulating a Sense of Place', *Australian Art Education*, J. Aust. Inst. Art Education, vol. 15, no. 2, pp. 20–3.

THEME IV

Making: creating gardens and the evolution of styles

Every young man who wears a gold ring in one ear is making a statement about his values and his attitude towards society. So it is with gardens. Gardens are statements – or 'signs' in semiotic terminology – and in any case, the carriers of meaning. Their meaning, however, is subject to interpretation, as with the young man's earring. He may wear it to be a nonconformist, rejecting the unadorned image of the male in his father's generation – but he may equally be conformist, in doing just what all his peers do. Nor is his own account privileged: he may be deluded, or lying, or lacking insight into his own motives. This is again true of the interpretation of the statements made by gardens, to which much current literary theory is applicable. There are as many 'readings' of a literary work as  there are readers, and none is privileged. Nevertheless, some are a lot more interesting than others.

The readings I give of gardens, some of them my own gardens, are also my own and only mine. I do not offer them as 'the Truth', but as the self-exploration of a designer – and we are all designers in our own gardens – to help us to think about the kind of statements we are making. These may often be driven by fashion, or be unexamined reflections of our native culture, and in that sense the statements are ventriloquial: others are speaking through us. There is nothing wrong with this, but it is nevertheless better to be aware that we are acting as a ventriloquist's dummy. The current vogue for white gardens and silver gardens and gardens divided into rooms, for example, represents in large part the voice of Vita Sackville-West and her garden at Sissinghurst.

Some primary garden types are derived from our evolutionary history, beginning with the garden as a forest clearing and the garden as an oasis. The contrived pastoral landscape and the Arcadian landscape dominated Australian painting for many years. The 'heath garden' has been influential in Australia, where many 'native' gardens are essentially heath gardens; others are 'bush gardens', mimicking the natural scene. However, although such gardens arose partly as a response to functional needs, especially in water conservation, they create functional problems of their own, not yet fully resolved. So some discussion of functional problems is also necessary.

In considering the above, we need to explore attitudes to order and control, as against apparent disorder, the laid-back and casual. A closely related set of themes springs from the preference for symmetry or for asymmetry, and the different power structures each supports (both of these themes were introduced at the beginning of this book, in Chapter 1 and Chapter 2). A third set, also closely related, is the naturalistic–contrived pair. Attitudes to boundaries are also critical; gardens can be regarded as enclosures (keeping something in) or as exclosures (keeping something out) or both. Boundaries can also be dense and impervious, or fairly open and pervious. This is true, moreover, of both space and time, a point made most elegantly by Chen Congzhou (1990, p. 5): pervious boundaries in space allow 'scenery outside gardens' – the garden depends on 'borrowing' the external landscape. Temporal boundaries that are pervious allow 'scenery outside scenery', and so the garden depends on 'time' – that is, a garden can be confined narrowly within a contemporary temporal framework, or it can be rich in historical associations, sometimes formal, often literary in China, such as inscribed couplets on stone or wood, which carry the contemplative back through a long cultural tradition to an ancestral past. This concept is also found in the West, but in post-modernist times the past has become a kind of convenience store, from whose shelves we snatch arbitrary and unrelated items as they take our fancy. That 'less' can be 'more' is a lesson hard to learn in the supermarket of historical and floristic superabundance, and I have struggled to learn it myself. The choices we make on all these matters reveal a great deal – make statements – about our individual personality and our geographic and social context.

The instinct to garden and the human needs met by gardens may be virtually universal, but the word 'garden', and its equivalent in other languages, has no single meaning, and in that sense, there is no universal garden, nor is there likely to be, if we think in terms of garden form. One fundamental distinction is that between the word used as a part of the language of 'the high culture', and the same word used as part of the popular culture. The two intergrade in minor degree in the English-speaking world, but for the most part, the historic origins, the descriptive language, the evaluative criteria and the socio-economic status are different. Discussion of gardens like Sissinghurst in Kent, in south-eastern England, or the Villa d'Este in Italy, or the great gardens of Suzhou in China, or for that matter contemporary gardens like those of Burle Marx in Latin America, are described and evaluated in terms of 'the high culture'.

In Australia, however, gardening is without question also one of the popular arts. It has the highest ranking of all recreational activities. It is big business for the horticultural and garden supplies industries. It is practised almost universally: Australia and New Zealand have the highest home-ownership in the world, and 'home' means a house, often a single storey, on its own block of land, with room for a substantial garden both in front of and behind the house, and narrower garden strips on each side. There is reason to think that the common association of the garden with the private house was prefigured by the Romans, although their private garden would have been restricted to the affluent in a slave society, and not self-maintained. When Australians talk about 'the garden', they mean their *own* garden, they have planned and designed it themselves, and they do nearly all the work in it themselves, including composting, pruning, spraying, and mowing the lawn – with the help of a battery of power tools and other equipment. They generally also lay their own reticulation systems, to water the garden, with an electronic control box.

I make the point because such a situation is inconceivable in much of the world. It is largely restricted to Australasia, North America and northern Europe. If every Chinese household were to have a garden the size of the average Australian suburban garden, together with the land needed for roads, rail and necessary open space, we would need some 80 million hectares, an area three times the size of Italy, just for private gardens. Nevertheless, gardens springing from the popular culture make up the greater part of the world's 'ornamental' gardens – that is, gardens whose function is decorative or pleasure-giving rather than materially productive.

An interest in the gardening culture of Australia has only just become reputable academically, as part of the broader study of popular culture, yet it has long seemed to me to offer a good entry point for looking at environmental and other cultural values. The garden is one of the very few places where the individual can improve on the handiwork of the Creator, who may have got off to a good start with the Garden of Eden, but hasn't done much since, certainly not in our suburban block, where we have had to start virtually from scratch. In this section, Chapter 16, The suburban garden in Australia, is a short essay on the suburban garden – mostly the front garden – and Chapter 17, The Australian backyard, is a complementary piece.

The third essay, Chapter 18, Gardening across Australia, is an account of my own evolution as a gardener. I would like also to have included a sample from *Swan River Landscapes*, published in 1970. It has a good slogan, well before its time: *Fear the hose*. A quarter of a century later we talk about 'water-sensitive design', but the case I put in *Swan River Landscapes* is made on aesthetic as well as conservation grounds. Indeed, the book is full of aesthetic criteria, to some extent bolstered by invoking the *genius loci*, Wordsworth, the New Testament, indeed anything that came to hand, since it was intended to be persuasive. I still hold the same aesthetic views and still make no apology for them (they are *right*, dammit). At least I argued for them, put a case, tried to make my criteria explicit, in that book. What I do not do is to try to trace their source in my own temperament and background. That is what I try to do, so far as I can, in Chapter 18, which follows me through five gardens, differing in space and time. I have tried to do the same for *Sense of Place* elsewhere in this collection (see Theme III). Australian writing in general is much preoccupied with place. Does this indicate a secure relationship, an insecure one, or both, or neither? Is the intense commitment to our own private and personal spaces (our gardens) a part of the same story?

Chapter 19, The garden as Paradise, turns from the popular to the gardeners' gardens, the work of a large group who take gardening seriously and critically, attend conferences on gardening, read widely, take garden tours overseas. This group is also a significant feature of Australian life, generally well educated, well travelled, and highly conscious of the 'inescapable tensions of being Australian', from the attempt either to adapt admired models from other lands to Australian conditions, or, increasingly to seek alternatives, still evolving. 'The Garden as Paradise' was written

147

for the October 1996 conference of the Australian Garden History Society, where I was asked to provide notes on the cultural history of some key terms.

References
Seddon, G., 1970, *Swan River Landscapes*, Perth: University of Western Australia Press.
Congzhou, Chen, 1990, *Of Chinese Gardens*, Shanghai: Tongji University Press.

16 The suburban garden in Australia*

Beginnings

There is a great variety of suburban gardens in Australia, but there are some common characteristics. Typically, the suburban garden is *negative space*, what falls between two very positive boundaries. One is the boundary of the block, looking out; the other is the wall of the house, looking in. The 'garden' is what is left between, a kind of no-man's-land, scarred by earlier events. The block has been surveyed, the bulldozers have done their work, cleared away the trees and topsoil, the builder has come and gone, leaving a rubble of plastic odds and ends, broken tiles, dropped nails, shattered concrete. This is the birth of the suburban garden.

There are other ways of designing and building, but they are rare in Australia. One is to survey the site, orient the proposed building in relation to the topography, keep the trees, whether indigenous or from an earlier garden, and maintain the landform, rather than flatten it. All of this happens at times in relatively affluent new developments in forested areas like parts of Sydney's North Shore, parts of the Dandenongs, the Loftys and other hill stations on the urban periphery. But it costs the developer more, and requires care, so it is not common practice. Another way to build is to take the block of land itself as the design unit – a common example is that of the building as boundary, with an internal courtyard, so familiar in Europe and Latin America. But our building regulations do not encourage this, although it often suits our climate. So we buy a block, flatten it, and plonk a house in the middle.

Why? The answer lies primarily in the way our cities have been surveyed and regulated, but behind these codifications lies a set of attitudes, substantially English in their origins, and far from simple. Dickens catches the territorial imperative of the suburban English, that intensified sense of a defensive space between house and the external world, in *Great Expectations*. Pip visits Wemmick in Walworth, in nineteenth-century suburban London.

*Excerpted from an essay first published as 'The Suburban Garden in Australia', *Westerly*, vol. 35, no. 4, 1990, pp. 5–13.

Wemmick's house was a little wooden cottage in the midst of plots of garden, and the top of it was cut out and painted like a battery mounted with guns. 'My own doing', said Wemmick. 'Looks pretty; don't it?'

I highly commended it. I think it was the smallest house I ever saw: the queerest gothic windows (by far the greater part of them sham), and a gothic door, almost too small to get in at.

'That's a real flagstaff, you see', said Wemmick, 'and on Sundays I run up a real flag. Then look here. After I have crossed this bridge, I hoist it up – so – and cut off the communication'. The bridge was a plank, and it crossed a chasm about four feet wide and two deep. But it was very pleasant to see the pride with which he hoisted it up, and made it fast; smiling as he did so, with a relish, and not merely mechanically.

'. . . At the back, there's a pig and there are fowls and rabbits; then I knock together my own little frame, you see, and grow cucumbers; and you'll judge at supper what sort of salad I can raise. So, sir,' said Wemmick, smiling again, but seriously, too, as he shook his head, 'if you can suppose the little place besieged, it would hold out a devil of a time in point of provisions.' . . .

'I am my own engineer, and my own carpenter, and my own plumber, and my own gardener, and my own Jack of all Trades,' said Wemmick, in acknowledging my compliments. 'Well, it's a good thing, you know. It brushes the Newgate cobwebs away.' (pp. 230–2)

The house and garden are a parody of that popular saying, an Englishman's home is his castle, complete even to a tiny drawbridge. London was Europe's greatest metropolis, and English cities were the first to industrialise, which both intensified slum crowding and created the affluence which offered an alternative to all but the poorest, by building detached and semi-detached houses, each within its little plot of land, in the new suburbs. The acute territoriality is a response to crowding. But Dickens' fantasy shows two other aspects of English social life. Everything in Wemmick's abode is scaled down, not something created in itself, but a minute version of something grander. This may be seen, perhaps, as illustrating the nature of English class structure, in which the aspirations of even the lower of the lower middle class are modelled on those of their superiors. A robust peasant and labouring class has its own cultural values, but these were eroded in southern England by the pervading force of the middle classes. Finally, Wemmick is Robinson Crusoe (a very English figure), dreaming the dream of self-sufficiency, using his *back* garden productively. This too may be in part a response to the insecurity of urban life in the first half of the nineteenth century.

Much of our background comes from Wemmick's Walworth: the omni-presence of middle-class values, the scaling down, the productive use of the back-yard and its sharp differentiation from the front yard, and above all, that strong consciousness of the individual boundary, not nearly so common in other cultures: the Americans rarely fence their front yard, even at the sides; while most Europeans live, as they always have, in apartment buildings with a shared internal courtyard in a building complex often fronting directly on to a piazza or other public open space. The individual has some rights in each of these spaces, but different ones, and so the 'drawbridge mentality' has less encouragement in a milieu that offers a gradation from private to communal spaces.

The cadastral survey

Another major cultural force that underwrites suburbia and its gardens in Australia is the cadastral survey, that rectilinear grid imposed at various scales on an entire continent by the geometers of a remote imperial power in the nineteenth century, squaring off this old, irregular landscape to impose an order convenient to an authoritarian colonial administration. It is so much a part of our lives that many Australians are scarcely aware of it as a massive imposition, yet a glance at the map shows only two natural boundaries at state level, the Murray River, and the Tasmanian coastline. The survey grid is a characteristic of empire: the Greeks used it in their colonies in southern Italy, and the Romans carried it throughout their imperial domain, ruler straight. The grid was usually set to the positions of the compass, the cardio (north–south) and decumanus (east–west) of the Romans – of which Hoddle Street and Victoria Parade in Melbourne are an example, set by Victoria's colonial surveyor as the axis of his grid nearly two millennia after the Romans. The British took it wherever they imposed their rule: Ireland, India, North America. It was convenient for the movement of troops, for making a census, for imposing taxes. It has the logic of a central administrative power, but it ignores natural features, and has often had a heavy cost in Australia, where, for example, most farm fences now turn out to be in the wrong place. There are a few early subdivisions which run to a different logic – the narrow river frontages and long narrow blocks running back from them at Guildford, also at Hahndorf in South Australia. Much of early Sydney has an irregularity of form that is the product of a topography that could not be ignored because it was so rocky and so steep, but also because there was an improvised and adaptive quality to its first phase of settlement.

But regularity was the rule in suburbia until the curves of Canberra became popular. Thus the legendary 'quarter-acre block'. This notion is now a part of our self-image but we might note that although the quarter-acre block was fairly common, it was never standard, either in Sydney or elsewhere in Australia – there was great variation in size both in and between cities. The 'classic' quarter-acre was 100 links (or one chain) by 250 links (or two and a half chains), with a frontage of 66 feet and a depth of 165 feet (roughly 20 metres x 50 metres, or 1000 square metres). This was the standard subdivision in Dalkeith and Nedlands in Western Australia, for example, but blocks in South Perth were subdivided at the same time (1916 onwards) at 32 perches, which represented four-fifths of the Dalkeith standard. In Claremont, an older suburb than Dalkeith, frontages were generally narrower (often 55 feet) but many blocks were deeper, 180 or 200 feet. The critical point, however, is that most suburban houses everywhere in Australia other than the innermost cities had big backyard spaces by world standards.

The measurements and their names – chains, links, perches – are quaint to those brought up with the decimal system, but they were critical in the production of Australian suburbia, with all the values and design habits this has engendered. The measurements have an ancient and a practical origin. An acre was the land that one man could plough with one horse in one day; as a strip of cultivated land, it could conveniently be measured out by two chains in width, five chains in length. The system lasted because it was familiar and convenient. The 'chain' became valued culturally because it was the length of a cricket pitch, but it has a major practical value in being both Imperial and metric, 66 feet, but 100 links. In short, it was a magic number. Roads, for example, were commonly one chain wide. When they curved, the width had to vary. The surveyor could then use trigonometrical tables to calculate the variation from 100 links by shifting a decimal point. It was a physical chain until well into this century – first the Gunter chain, then a steel band. Many country towns, especially in New South Wales, were laid out in 10-chain square blocks, and then subdivided into housing blocks that were one chain wide and five chains deep, half an acre, and very deep (for example, Tumut). The system began to be metricised in the eastern states from the 1920s – the ACT, for example, used a metric foot – but it persisted in Western Australia until the introduction of the metric system in the late 1960s. So many of us now live, in a city, on a block of land of the length that one man with a horse could plough a straight furrow, and the area that one man with one horse could plough in one-quarter of a day.

The form remains, with all its virtues and its costs. Intellectuals in Australia sometimes under-rate the virtues, which are very real – the spaciousness, the privacy, the room for children to play and for adults to entertain, room for the pool, and room to be Wemmick, your own engineer and carpenter and gardener. All of this is seen as a great privilege by the crowded urban dwellers of less-happier lands. The costs are also great: above all the costs of servicing such a low density, which makes public transport grossly uneconomic and thus leads to car-dependence, with all its problems, including the isolation of young women with children, and of the elderly and of young teenagers – all those *without* a car become locked in their green prison.

The aesthetic cost is also great. In a typical quarter-acre block suburb, the natural diversity of landform, soil, vegetation and aspect has all been wiped out and the grid then emphasises the monotony. But we crave diversity, so that a spurious diversity has been imposed, lacking any natural logic. And so that which should be diverse has been made uniform, while that which should be, if not uniform, at least homogeneous – the built form and the continuity of landscape style – has become heterogeneous.

152

References
Dickens, Charles, (1861) 1958, *Great Expectations*, London: Longman, Green.

17 The Australian backyard*

Functions of the backyard

When I was young, in the 1930s, and for several decades on either side, the function of the typical Australian backyard in the cities and country towns could be known easily from a list of its contents. It had all or most of: a woodheap, often with a rickety woodshed with a low roof of galvanised iron and a fence for the back wall; a wash-house, with two tubs and a copper, with a grate beneath it to heat the water and a wire rack to hold the Velvet soap and Reckitts Blue; a clothes line; one or more tanks on wooden tankstands, with mint and parsley under or near the dripping tap in a cut-down kerosene tin; a dunny against the back fence, so that the pan could be collected from the dunny lane through a trap-door; there might be a kennel for the dog, although he often slept under the verandah; there was sometimes a crude incinerator, often an old oil drum, although rubbish was also burnt in an open bonfire. There might be chooks, usually in a chook-house along the back fence, and sometimes a sleep-out, usually a verandah enclosed with fly-wire, but sometimes free standing. A lemon tree was nearly universal; other trees varied with climate – almond trees in Adelaide and Perth, plums and apples in Melbourne, choko vines and bananas in Sydney and Brisbane, a mango in Cairns, figs and loquats almost everywhere. For a few weeks, there was gross overabundance of fruit, and much trading ('I'll take some of your plums if you take some of my apples next month'). Blackbirds, Ceylon crows and starlings grew fat (except in Perth). They didn't mind the fruit fly grubs and codlin crawling in the apples. In the country towns, there was a good chance of a pepper tree (*Schinus molle*), which left a grubby latex film on your hands and clothes when you climbed it. In Kalgoorlie, where water came in a pipe from the Darling Scarp, the shower water was drained out to a banana plant. Sometimes there was a patch of coarse grass, couch or buffalo, for the kids to play on, but there was rarely any special provision for the young, who played under the

153

*First published as 'The Australian Back Yard', *Australian Garden History: Journal of the Australian Garden History Society*, vol. 3, no. 2, 1994, pp. 2–9, and reprinted in Ian Craven (ed.), *Australian Popular Culture*, Cambridge University Press.

A backyard shows its functions: woodheap, outhouse, wash-house, tankstand with tank and a bucket to catch the drip from the tap.
SHIBU DUTTA

tankstand, in the woodshed, in the back lane, or in the driveway – which was good for cricket. Swings and sandpits all came later, as did swimming pools and barbecues.

I don't think there was much regional variation, and not much change either, over a period of fifty years or more. Wooden-slatted fern and orchid houses were fairly common in Brisbane, and staghorn ferns were common even as far south as Melbourne. In Tasmania, the wash tubs might be made of Huon pine, in Victoria of concrete. Drier places like Mildura might have two tanks for rainwater rather than one. Some people had vegetable gardens – onions, peas, beans, cabbages, lettuce – and a compost heap to go with it. There was often junk piled up somewhere in the yard, since councils did not come round to collect it. From the late 1930s onwards, there was sometimes an old car body, enlisted by the young as play equipment. The advent of the car was a major change, adding the driveway, perhaps a garage, and a new activity (washing the car). In the last thirty years, changes have come thick and fast. The basic functions of the backyard have changed, but we will come to that later.

Not all backyards were the same, of course, but the variations were not so much regional as a reflection of differences in social standing and ethnic background. The Italians grew tomatoes, onions, oregano and tarragon, zucchini, fennel, olives and wine grapes, and sometimes 'rolled their own'. So did the Greeks, who also grew tomatoes, and two or three different kinds of basil, although any but the most common herbs (mint, parsley, thyme and sage) were hard to get in Australia before World War II, and the herb garden of today's fashionable middle-class suburban cook was unheard of. The Chinese, as always, cultivated every inch of ground available to them; Tom Hungerford (1977) gives a good account of such a garden in South Perth in the 1930s. Greeks and Italians to this day often grow vegetables in their front yard as well as their back, in inner-city suburbs like Richmond in Melbourne and Leichhardt in Sydney. The backyards of the German settlers in the Barossa were more orderly and better cared for than the Australian average, but not essentially different in function.

The vegetable garden probably showed the greatest variation. A raised bed was made for growing vegetables in many areas, but not all, reflecting the regional practice in the British Isles from which the settlers came. This neat raised bed, bordered with wood, brick or beer bottles gave good drainage. Narrow rectangles with gravel paths between gave good access. The bed itself was built up with compost, and mulched with straw and horse dung. I have seen survivors in Port Fairy in Victoria, an Irish town, and in Hobart, but the formal plots are not now so common; vegetable beds are still often mounded, but they rarely have raised borders, except where the permaculture system returns to old ways.

The other variation was social. Rich people had much bigger backyards, often with stables, a tennis court on the double block beside the house, more fruit trees, a bigger vegetable garden, a cutting garden for flowers for the house. Yet rich or poor, most of the domestic functions still had to be met.

Function is the key word. The backyard, equally in the town and the country, was complementary to the house in providing resources for living; storage, water, fuel, washing facilities, food input, and food output (by way of the dunny and the compost heap). It also served as a male domain, while the house was female. The women did the washing, and perhaps the flower garden if there was one, but the men chopped the wood, usually lit the copper on wash-day, and looked after the vegetables, washing up in the wash-house, as country men still do. The bathroom inside was for the women, and men continued to use the outhouse long after a toilet was installed inside. Glen Tomasetti faithfully records all these rituals and uses in *Thoroughly Decent People* (1976), in which she is describing East St Kilda in 1934.

> Schooled to the remote privacy of the outhouse, no sewerage would induce Bert to have or use a lavatory in a bathroom. He continued to use the wash house for shaving and scrubbing in preference to the bathroom which, unless he needed something from the medicine chest, he entered only twice a week to have a bath (p. 24).

> Bert was up at half-past six and Lizzie at seven. She didn't have tea in bed on Monday morning because it was washing day. Bert had filled the copper with water and, since he had lit the fire under it for so many years, he now lit the gas (p. 22).

> While the men and boys played cricket in the drive with a kerosene tin for stumps, a good old bat and a tennis ball, the girls went inside to help Lizzie get the tea (p. 4).

So the suburban backyard served as play space and, for imaginative children, a magic carpet that could become many things. But that is not what it was *for*. In being a necessary adjunct of the house serving domestic needs, it was essentially *rural*, a gesture towards functional self-sufficiency, not complete, but not totally dependent on a web of urban services as we are today. The suburban backyard was not fundamentally different from the country backyard in Australia. Indeed the main difference was only that the country backyard usually had more junk – because it was harder to get rid of. The suburban Australian backyard had no equivalent in any Italian city, or inner London or Dublin or New York or Tokyo, nor does it today. They never had the space, and the domestic functions had to be served in more compact ways. In London, there was no woodheap in the tiny 'area', as the space was called, but coal in the cellar, reached directly from the street. In Rome or Hong Kong, washing still hangs from upstairs windows. The cities that have something like the suburban Australian backyard are those that have grown in the last one hundred years or so, swelled by a migration of rural people to the city:

155

cities of countrymen and women. Some are rich, some are poor; Los Angeles is a rich example, but full of Dust Bowl farmers who walked off the land in the Depression. Port Moresby is an example of a poor city, made up of villagers who have to keep up a degree of self-sufficiency because urban services have not kept pace with their arrival, and who need and want to keep up a degree of self-sufficiency because they are culturally attuned to it: certain foodstuffs that must be fresh, or that cannot readily be obtained, or that should not be touched by other possibly malign hands. Most squatter settlements on the outskirts of Third World cities have these characteristics – because the squatters come from a culture of self-sufficiency, and because, in any case, there is no choice. Similar forces applied in our suburbs. The dream of almost every immigrant was to acquire land of his own, no matter how little, but production on the quarter-acre block was also a product of living conditions characterised by poor supplies, no refrigeration, indifferent urban services – and recurrent poverty and shortage, as in the depressions of the 1890s, the 1930s, and much of World War II.

The inner city

But if there is a functional continuum from country backyards to suburban back-yards on the quarter-acre subdivision, there is a break as we move in to the inner city. These areas have a different history, especially in Sydney and Melbourne – although they now represent only 2 per cent and 1 per cent respectively of the housing of those two cities, they represent a distinctive urban form and culture (Neutze, 1977, p. 5). They spring from rapid population growth in the second half of the nineteenth century.

In some inner-city areas, especially in inner Sydney, densities were very high. By 1891, the 3500 people of Darlington were housed in 672 five-to-six roomed houses built on 62 acres (25.1 hectares). The density per acre (0.4 hectares) was 61.88, but this is less than the density of habitation, because it takes no account of space used for shops and offices, and therefore not available as residential. Paddington had 44.11 persons per acre, and Redfern 46.86. Several wards of the City of Sydney had a high average of persons per inhabited dwelling. In the ward of Bourke, there were 8.2 persons per dwelling (often of only four rooms) in 1891. 'In Long's Lane, off Cumberland Street, seven houses shared one water tap' (Kelly, 1978, p. 74, from whom these data are drawn). The backyards were minuscule and filthy, with a water closet at the back door.

They were lucky. A court in the Rocks – Miller's Buildings – of fourteen houses, each of two rooms less than 3 square metres, had four closets for the fourteen houses, which were estimated to house about sixty people. In 1889, the investigating committee 'found on the doorstep, a heap of human excreta, covered with an old straw hat' (quoted in Kelly, 1978, p. 76). In 1900, 303 people contracted bubonic plague, and 103 of them died. This at last brought these conditions into the limelight, and led to slum clearance and better sanitation in inner Sydney.

In 1890, however, working-class housing in Sydney was thought by those few who had studied it, to be worse than that of London, which was generally agreed to be worse than anywhere else. Rapid growth in population and the loss of inner residential land to industrial and commercial use led to a growing population, trapped by poor public transport, crowded into a decreasing area, served by an incompetent local government, in a steep sandstone terrain that of itself made the provisions of adequate urban services difficult.

Water supply and sanitation had been difficult in Sydney from the outset: in 1851, only about 1000 houses of an approximate 8000 in the Sydney Corporation area were connected to mains supplies. Many houses in the 1850s had wells and cesspits side by side. A report in the *Sydney Morning Herald* said of the inhabitants of Paramatta Street that:

> they cook in dirt – they eat in dirt – and they sleep in it, they are born, bred and they die in dirt; from the cradle to the grave, they pass through life in filth – society tolerates it, and they look upon it as their inheritance. (1851, quoted in Clark, 1978, p. 57)

'Marvellous Melbourne' was little better than Sydney in the colonial years. Bernard Barrett (1971) has given a detailed account of the slums of Collingwood, and the uses to which backyards were put there.

> In the mid-nineteenth century privy out-houses were usually constructed over or near a cesspit. Cesspits were of varying degrees of sophistication. In the 1850s and 1860s the typical cesspit on Collingwood Flat tended to be at the primitive end of the scale – a mere hole dug in the ground. It was probably never emptied; when it became filled with solid matter, it might be covered over with earth, and the timber superstructure would be moved a few yards away to a new hole. (Barrett, 1971, p. 75, in a chapter with the title 'From Cesspits to Cesspans')

They grow good tomatoes in Collingwood backyards today.

The Board of Works was created in 1890, with the responsibility of providing water and developing a sewerage system. Melbourne had already established a clean, continuous and publicly owned water supply by 1853, while London's system did not reach this stage until 1899, but Melbourne was well behind in establishing a sewerage system. London had made cesspools illegal in 1847, although it then ran its sewers into the Thames. Adelaide began constructing a sewerage system in 1878, and Sydney shortly afterwards. Hobart and Melbourne constructed their major works in the first years of the twentieth century. The inner suburbs of Perth were sewered at a leisurely pace, as befits the more relaxed lifestyle in the west, between 1906 and 1920, with the main outlet at Claise Brook on the Swan in East Perth. 'Some wealthy households installed their own septic tanks. Most made do with the double-pan system and dry-earth closets until the sewerage pipes reached them' (Stannage, 1979, p. 278). Brisbane was the last of the capital cities, on a timetable like that of most country towns. The construction of Brisbane's first sewerage project began in 1916, but did not proceed until 1923, when 'pan closets were still operating in central Brisbane.

'They still operate (along with septic tanks) in some Australian country towns and outer suburbs today. In 1960, it was estimated that half a million people lacked mains sewerage in the Sydney metropolitan area alone' (Barrett, 1971, p. 137). Perth in 1988 was 73 per cent sewered; septic tanks in the Darling Ranges began causing concern, and led to an official *Report of the Select Committee Appointed to Enquire into Effluent Disposal in the Perth Metropolitan Region* (finally leading to action in 1996). Thus if one of the prime objectives of today's town planning is to connect bums to oceans, its success is not yet complete in Australian cities, and the backyard often continues to perform one of its most basic functions. The unsewered subdivisions today are not, however, in the inner city, which has different problems and prospects. They are in the outlying suburbs, especially those in the hill areas that adjoin all of our capital cities. These are expensive to service adequately, but suburbia continues its low-density invasion.

Rural origins of the suburban backyard

Australia is often described as one of the world's most urbanised nations, but this is misleading. For most of our history, most of us have been living in a suburb. Our culture still has a semi-rural flavour, although things are changing, and our backyards reflect it. In fact, our backyards faithfully reflect the history of the word 'yard', on which the *OED* has a long entry. First, there is a range of Teutonic words (OS. *gardo*, yard, farm, MHG. *garte*, G. *garten*, garden, Goth. *garda*, enclosure) and so on. It goes on to say that 'close affinity of sense is exhibited by the words derived from the Teutonic root . . .'. The primary sense is that of 'enclosure' – of which the circle is the most economical form, used by cattlemen from the Bantu to early Australian bush-drovers, whose roughly circular corrals can still be found in quiet decay south of the Monaro. The basic enclosure was either to keep cattle in, or keep them out. It was sometimes qualified by a prefix – farmyard, vineyard, orchard. In short, the word was used for a multifunctional enclosure, generally attached to a house, basically rural in origin. There is a similar set of words 'derived from an Indo-European root *ghort*, viz, Gr. Xopros farmyard, feeding place, food, fodder, L. *hortus* garden, *co-hors* enclosure, yard, pen for cattle and poultry – but there are phonological difficulties in the way of equating both groups of words'. Whether they can be equated or not, it is interesting that both sets of words have a similar range of meanings.

158

So there is something primitive about the Australian suburban backyard, both word and thing. Indeed the word 'yard' itself is faintly archaic, more used, at least unqualified, in North America and Australia than in the British Isles. The verb form, used as in 'yarding cattle', is given as '*Colonial* and *U.S.*', with a quotation from Charles Kingsley's novel, *Geoffrey Hamlyn*, 'Well, lad, suppose we yard these rams'. There are few rams to be yarded nowadays in St Kilda or Double Bay, few outhouses or woodheaps or chooks or coppers, and not many vegetable gardens, either, although the lemon tree seems to be assured of eternal life. Self-sufficiency is no longer desirable to most people.

Does the suburban backyard described above go back to founding days? Denis Winston, the foundation professor of town and country planning at the University of Sydney until his retirement in 1974, emphasised both space and function as follows:

> With wide streets went large building plots; even the town-lands in Adelaide had originally one acre plots: horses, cattle, hens and pigs had to be provided for so that good yard space and extensive out-buildings were general. Even today Australians expect that a family home should accommodate the two cars, with trailer or caravan, and have room for the children's tent as a summer sleepout; and many home sites relatively close to the centres of the main cities are still big enough for this. (Winston, 1976, p. 188)

In Perth, there was a fine debate in Council in 1876 on pigs. The Medical Officer thought they were injurious to health and the Attorney-General proposed that no one should keep a pig within 50 yards (46 metres) of his neighbour's house – which would have had the effect of allowing them on the large blocks of the wealthy while forbidding them on the small blocks of the poor – but George Shenton and James Lee-Steere defended the poor man's right to keep a pig, and they won the day (Stannage, p. 174). Pigs were finally banned for ever in 1886 as insanitary, to be replaced by far more insanitary rubbish tips. This was part of the move towards centralised services that has characterised the growth of cities everywhere.

Changes

It is this that has deprived the backyard of its utility, or, more accurately, changed its functions. First, the coming of sewerage, then the advent of the motor car brought significant change; the car required a garage, also used for storage. There was sometimes an entry from the back lane (the dunny lane) into the garage in the back corner; if not, there was a long drive from the front, eventually paved, usually with two concrete strips and a well-trimmed grass median. The garage was usually behind the house. Later it grew in size, to accommodate two or three cars (and a trailer, and a boat, in homes that are more affluent but not necessarily wealthy); and it moved forward, flush with the frontage, of which it became an integral part, rather than an afterthought.

In the late 1940s and 1950s, the old clothes line went and was almost universally replaced by a horizontal windmill of steel and galvanised wire, known as the Hills Clothes Hoist (Hills began production in 1945); later, electric driers and retractable washing lines replaced the Hills Clothes Hoist in design-conscious back-yards – or, rather, back gardens, because this is what they were becoming. I asked a middle-class, middle-aged English friend to describe the backyard of her childhood home in Surrey; she replied, with mild affront, that they did not have a backyard, they had a garden behind the house. This was a class distinction that applied in Australia also to a degree, but as the nation has become more uniformly suburban and middle class, the distinction is blurred. Perhaps it is more accurate to say that people make statements of various kinds by the way they use the land around the house. Especially in country towns, the old backyard lives on. Others still use their backyard as a functional space, still primarily male, for working on the boat, maintaining the vehicle, fixing the trailer, stripping the paint off doors and mantel-pieces – it is still a service yard, although the services have changed. The atmosphere is casual, male, untidy, relaxed, spontaneous and, in its way, creative. It is emphatic-ally for use, not for display – and it is still common. I have seen some prime examples in South Fremantle this summer (1989).

159

The backyard as display

However, the trend is all the other way. The backyard has become back garden, for recreation, adult-dominated family use, and for showing off to one's peers. The following advertisement (from the Real Estate advertisements, *Sunday Times* (Perth), 18 December 1988) is typical of middle-range homes being offered in the new suburbs. It is *not* in the exclusive, luxury class of Claremont or Peppermint Grove, but is able to offer many of the same 'features', because the land is cheap.

> At the rear of the home there is a large and shady patio area, complete with gas barbecue. The rear garden is terraced and leads up to a paved area and a sparkling free-form swimming pool.

The area is beautifully landscaped with palms and shrubs, being sheltered by a shade covered pergola.

In becoming display space (entertainment), the backyard has added a public function to its private one, and thus acquired a characteristic of the front garden. We have not looked at that yet, but front and back are a dialectical pair, defining each other negatively, and to understand either, we must look at both. Once again, Glen Tomasetti sets the background with her Bert and Lizzie in East St Kilda, 1934. First, what Robin Boyd called 'arboriphobia':

> An enemy on one boundary, inoffensive people on another and friends at a short distance was a pattern repeated in the suburbs. The Larkins fulfilled the requirements for enmity. Their garden was neglected. Their flowering gums, planted right on the fence, dropped leaves and nuts on Bert's drive. Hanging low after rain, they wet his head when he parked the Vauxhall beside the house. (pp. 7, 8)

With this enmity to trees goes a mania for pruning, which is still alive and well.

> Arthur didn't believe in pruning soft-fruit trees. Bert did. He loved pruning, cutting back and lopping. He often walked round the garden working a pair of secateurs in his right hand, looking for dead flower heads and wayward twigs. When the day came to prune a tree, he started the job joyfully, cutting back to the last possible spot from which new growth might shoot. The sight of a tree, just after he'd pruned it, was as painful to Lizzie as the sight of her was to him, after she'd had a new permanent wave. (p. 4)

By implication, Tomasetti later attributes both these behaviour patterns to a pioneering mentality by showing Bert's reaction to natural bushland.

> They were passing through bush and it depressed him. He could see no beauty in it, no beauty at all. It represented only back-breaking labour. He thought of fire because he'd really like to put a match to it and see it swept away, leaving the land for man's use. That didn't happen after a fire of course. The bush recovered. The grey-green leaves of the gum trees with their ragged bark and the spindly wattles not yet in flower all filled him with dull melancholy; work, monotony, work. The bush had nothing to do with Bert's understanding of the glories of nature. (pp. 130–1)

This hatred of trees is still common. The following letter from a suburban newspaper (*Weekly Post* (Subiaco), 10 January 1989, p. 6) is not unusual:

> Those damn box trees! I must admit I call them stronger things than that. Well, I've just spent another one and a half hours raking leaves. When are the Subiaco council going to do something about them? The streets and footpaths are an absolute disgrace.
> We know the council has a thing about cutting down trees but why not revert back to pruning them round and every year but on both sides of the street? Surely in the long run it would be cheaper than the major job it will be one day.
> I thought I would try if you can't beat them join them, but we like a tidy yard and a drive and the road sweeper would need to come every day to sweep them away. Subiaco and Shenton Park would certainly never win a tidy city or street award.

Mechanical sweepers, which suck up leaves from roads, footpaths and grass, are common in Perth, used by councils, public institutions and home-owners. Although the hungry sands of the Perth metropolitan area are notably deficient in organic matter, the leaves so collected are rarely composted, even in educational institutions.

Tidy towns

A passion for neatness is the most striking characteristic of Australian display gardening, either institutional or private (see colour illustration following p. 174). Edges are trimmed, leaves are raked, flowers are staked, concrete hosed down, shrubs trimmed and clipped, trees pruned. The bounteous, brimming, rambling, over-blown, careless garden is still rare – with reason, in that it is actually harder to maintain in Australia, where growth is rapid, and the overgrown garden soon becomes no garden. But the driving forces behind this mania for neatness are highly conjectural. It is not peculiar to Australia – it is to be found in New Zealand, parts of the north of England, parts of Canada, less so in the USA. It may in part represent the pioneering spirit, but it is clearly also a cultural inheritance, as its distribution shows. There is a strong component of what is variously described as 'keeping up with the Jones's', 'peer group pressure' or 'civic pride'. In the pre-World War II Australia, ordinary, decent people kept up appearances, not without considerable effort, to maintain their self-respect; freshly ironed shirts for the schoolchildren and a tidy front yard kept the flag flying.

Perhaps this concern with appearances, although a common human character-istic, was intensified in Australia by the awareness of the convict background. Respectability had to be fought for. Popular phrases reflect a fear of 'going under', or 'giving up', of *failing* 'to keep the flag flying'. Outward signs of disorder might signal the rule of chaos and old night come again. The working classes of industrial Britain must have shared these fears. Rural immigrants and the unskilled poor these growing cities attracted and bred could so easily be forced by sheer circumstance into what were called vice and crime, which began a nightmare of subhuman deprivation. Scrubbing and sweeping and raking may have helped to keep these fears at bay.

But the mania for tidiness also represents a discourse with the environment. Our house and its immediate surrounds are one of the few areas in our lives where we have real power, make decisions, and put them into practice. Perhaps it is this sense of control, especially among those who have limited control of their own lives in the outside world, that leads Bert to prowl the garden, secateurs in hand, looking for something to 'manage' by imposing his will. If there is a distinctively Australian component in this behaviour, which on the whole I doubt, it is a rejection of the endless leaf litter and the asymmetry – the *untidiness* – of the Australian bush. Per-haps that fear of the red-back spider lurking under the toilet seat, waiting to strike when your defences are down, was seen as a malign outsider from the bush, a fifth column from that harsh natural Australia we have fought to control or exclude – since the fear was quite out of proportion to the occurrence.

Since it was always semi-public, a place to work in, but not for recreation or living, the front garden has changed less than the backyard, but it too has been subject to the vagaries of fashion. Some of the changes are shown in this last quotation from Glen Tomasetti:

'You look like having a good show of dahlias,' said Keith. 'They're always good after they've been lifted and you can stake them as you re-plant. You ought to have more dahlias instead of all those annuals and borders you plant: not worth the work.

'Valerie likes them for picking,' said Keith solemnly. He wasn't interested in gardening but did what was expected of him, only declaring his independence by growing different plants from those favoured by the family into which he'd married. Instead of hydrangeas, standard roses, dahlias, snapdragons, a lemon tree and stag ferns for the shady side, he'd tried lupins, delphiniums, hollyhocks, begonias and his lawns were pure English grass, fine and soft. He'd just bought two azaleas for the south side of the house and he knew his father-in-law would disapprove of this unnecessary expenditure. Bert, the son of a gold-miner turned farmer, could spend happily only on essentials. When boots or chairs or a pocket-knife had to be bought, he went for quality because it was economy in the long run. But gardens were to be made from other gardens, from cuttings, corms, bulbs and roots of buffalo grass. You had to buy a lemon tree, standard roses and grass seed for bare patches. (pp. 10, 11)

The azaleas and English lawn grasses mark the availability of a more abundant water supply, and a more reckless attitude to its use, with hose and sprinkler rather than bucket and watering can. The disciplines of scarcity are relaxed. Not only water is abundant and used wastefully. Fertilisers, pressure-pack sprays, pelleted snail-killer, all add to the convenience of gardening, as take-away foods, full of fat, sugar and salt, add to the convenience of eating. Gardening has become a conspicuous element in the consumer society.

The nurseries reflect this growth in consumerism, and promote it. They are now rarely places where embryo plants are nursed into being, but supermarkets which sell flowers – these showy sexual organs attract the customers, and the living organism from which they come is incidental. To turn the pot over to tap out the plant and so inspect its roots – the most important part of what you are buying – is 'unacceptable' behaviour.

A recent book asserts the following:

The gradual infiltration of native plants into our suburban gardens, and the corresponding withdrawal of European-styled gardens, suggests that a process of legitimation is being acted out that mirrors positive changes in the Australian's relation with the landscape. Certainly the low-maintenance factor recommends the native garden to the house-proud, but that recovery of leisure – the delivery from the garden's tyrannical domination of the weekend – also signifies a growing sense of accommodation with the land, through which culture and nature have been made to co-exist more harmoniously. While creative and adaptive, the highly stylised character of the versions of 'nature' found in the native garden is nevertheless controlled by an edging of old railway sleepers and a covering of woodchips, the latter suggesting how ambiguous the putatively harmonious relationship can be. (Fiske, Hodge and Turner, 1987, p. 30)

It is hard to know how to tackle assertions such as this, which are common enough in popular journalism. What is a European-styled garden? Is the 'style' of a garden necessarily dependent on the choice of plant material? Are gardens using Australian plants necessarily low maintenance? Why should an increase in leisure signify a growing sense of accommodation with the land? (Its most obvious outcome has been increasing pressure on natural resources.) How many 'native' gardens are either creative or adaptive? As for the 'old railway sleepers', most are new – cut for the garden trade. Ancient red-gum forests along our inland rivers are steadily and relentlessly felled to supply them. Is this harmony with the land?

I would myself like to see more use of local plant material, gardens that provide habitat for birds, gardens in which plants are chosen and sited with care so that they can thrive without excessive cosseting. Such gardens are rare, as they always have been. They are not generally a feature of popular culture today, except perhaps in the vegetable garden, and they were probably more common when manure was recovered in the wake of the milkman's horse, and water was too precious to be wasted.

So both back and front changed as the times changed.

The retreat to the backyard

As Rob Ingram put it:

> With seven-and-a-half million square kilometres of playground available, Australians have largely withdrawn to their own quarter acre.
>
> And why not? It's summertime and the livin' is easy . . . with the pool, the barbecue, and the old redwood setting. The suburban backyard has become the resort that we used to drag the caravan down the coast to. (*Sun-Herald* (Sydney), 1 January 1989, p. 116)

This retreat to the backyard – or 'patio' behind the inner-city terraces – doubtless reflects attitudes to congestion on the roads and overcrowding and high prices at the resorts. Despite the transformations in function, the standard subdivisions persist in their millions, and thus the backyard space itself remains, or has done so until very recently. The free-standing house with space before and behind has been the Australian dream, but that too is now changing. Units, strata titles, duplexes, apartments, row housing, infill housing, penthouses, townhouses, all are names for denser living with reduced outdoor spaces. Australia, for better or worse, or both, is becoming urban, and the generous old backyard may become a threatened species.

Gardening is still a passion with many Australians, and although our cities are changing in so many ways, it is hard to doubt that they will not continue to garden, provided they still have a place to do it. If we lose it, as seems likely, we shall have to fight for more human, more usable, public open space.

163

References

Barrett, Bernard, 1971, 'The Inner Suburbs: The Evolution of an Industrial Area', Melbourne University Press.

Clark, David, 1978, '"Worse than Physic": Sydney's Water Supply, 1788–1888', in Max Kelly (ed.), 1978, *Nineteenth-century Sydney: Essays in Urban History*, Sydney University Press.

Fiske, J., Hodge, B., and Turner G., 1987, *Myths of Oz: Reading Australian Popular Culture*, Sydney: Allen & Unwin.

Hungerford, T. A. G., 1977, *Wong Chu and the Queen's Letterbox: The First Collection of Stories*, Fremantle Arts Centre Press.

Kelly, Max (ed.), 1978, *Nineteenth-century Sydney: Essays in Urban History*, Sydney University Press.

Neutze, Max, 1977, *Urban Development in Australia*, Sydney: Allen & Unwin.

Stannage, C. T., 1979, *The People of Perth*, Perth City Council.

Tomasetti, Glen, 1976, *Thoroughly Decent People*, Melbourne: McPhee Gribble.

Winston, Denis, 1976, 'Nineteenth Century Sources of Twentieth Century Theories, 1800–1939', in Seddon, George, and Davis, Mari (eds), 1976, *Man and Landscape in Australia*, Canberra: Australian Government Publishing Service.

18 Gardening across Australia*

Perhaps the most important benefit of gardening is the freedom to establish and define one's own territory, vital for many animals in creating confidence in the security of the immediate environment. In our own garden we can make our own decisions, give free rein to our creative impulses, make our own mistakes, and learn from them, change our minds, watch and observe the consequences of our actions, gain some insight into natural processes, and tie ourselves to the rhythm of the seasons. I don't see how a real gardener can lose interest in life: we always want to see the wisteria bloom again in spring, or in my case, see the wonderful orange-yellow flowering cones of *Banksia praemorsa*, and the little nectar-feeding birds that share my delight in it.

These pleasures are personal ones. You have chosen the plants, put them in the soil, cared for them; they are dependent on you, and their growth is your reward. The owners of a great estate with a garden professionally designed and maintained may take pleasure in their garden, too, but the pleasure will be of a different order.

I have made five gardens over the last forty years, first in Perth, then Sydney, then two in Melbourne, and then again in Perth. Each is different, partly because of location and climate, partly because of available resources, and partly because of changes in myself. One theme that runs through all of them is that gardens can be experimental. Another is that we learn from books, travel, and other gardens, and these generate ideas that we may like to try for ourselves. But I think that as we grow older, we also become more reflective about the influences that have fashioned our own tastes. Even when we think we are being most original, especially when young, we are often guided by our cultural background, and accept some courses of action as 'right', quite unthinkingly. How could it be otherwise? It takes experience and self-knowledge to learn that it *could* indeed be otherwise – and I shall give some examples from my own life. Yet another theme is that gardens can be resonant with associations of all kinds, and one's own knowledge can enrich these. Again, a

*First published as 'The Evolution of a Gardener: part 1', *Landscape Australia*, vol. 17, no. 4, 1995, pp. 279–82, and 'The Evolution of a Gardener: Fremantle: part 2', *Landscape Australia*, vol. 18, no. 2, 1996, pp. 43–9.

personal example: I have had the good fortune to travel extensively. It has given me intense pleasure to see some of our garden plants in their natural setting: in China, I have seen camellias in Yunnan Province growing naturally, and *Anemone hupehensis* along the roadside in central China, lacecap hydrangeas again along the roadside in western Szechuan, and *Magnolia liliflora* on the slopes of Fuji in Japan, *Arbutus unedo* in the *maquis* of Provence and the Mediterranean littoral, with *Arbutus andrachne* in Greece, *Arbutus menziesii* in Oregon, along with *Mahonia aquifolia*, or the superb *Protea* in the Drakensberg in Natal, in southern Africa. The pleasure and excitement of seeing these plants in their natural setting is redoubled on seeing them again in my garden. The two reinforce each other, and of course, one learns something of the ecology of the plant by seeing it in its native haunts.

But I like to know where all my plants come from, as do most dedicated gardeners, and also to know their history of cultivation, and their nursery experience also. A good knowledge of plants gives us a very concrete sense of the real world – there are few countries with which we are not able to associate a number of plants, even if we have not been there. Chile, for example, brings to mind for an Australian, a list of plants that have a Gondwanan origin: the monkey-puzzle tree, for example, is *Araucaria araucana* – and closely related to our Norfolk Island pine, *Araucaria heterophylla*, of which I have two in my garden – or the vivid fire tree, *Embothrium coccineum*, a close relative of our banksias and waratah, which it resembles, and the proteas of South Africa, all Gondwanan. And so on. I could go on indefinitely. But the point is that such knowledge is intensely personal, and personally enriching. In one's own garden, one is in contact with the whole globe, both cognitively and imaginatively.

Claremont in Perth

My five gardens, have all been different; the first in Claremont, a suburb of Perth, Western Australia; then Paddington in Sydney; then Hawthorn in Melbourne, followed by Richmond, also in Melbourne, and finally Fremantle, again in Western Australia.

My *aims* have evolved through time to:
- improve ecological fit;
- meet functional requirements, which have varied through time and space – for example, play space for our children when young, a place to dry laundry, maximum summer shade in Perth, maximum winter sun in Melbourne;
- express my 'sense of place' – to design a garden that fits well with both the natural and the social environment. However, these were often in conflict. For instance, when we began our first garden in Perth, our neighbours hated trees, but we loved trees and planted many suited to the soil and climate. They fitted well with the physical environment, but not at all with our neighbours' gardens.
- meet my aesthetic criteria. These have also changed with time and experience. Thus the gardens are a journey and a personal evolution.

The first garden I made, in Claremont, was around a house built in 1890 on a block a little narrower and a lot deeper than the 60-foot (18-metre) wide block that became common a few years later. This was the first house we owned. In retrospect, I find three things of interest about that garden. The first is that we planted trees, lots of them, at a time when most Perth gardeners avoided trees, or at most had one or two. Trees were messy; they dropped leaves everywhere, which had to be raked up, they blocked the gutters, which had to be cleaned, many of them had gum, flowers or exudates that marked the roof of the car and stained bricks and paving, and, above all, they shaded the roses and displays of annuals, while their roots robbed the nutrient from all around them.

We didn't care. We loved trees, and were early members of the Tree Society, which vigorously fought the current arboriphobia. So we planted lemon-scented gum (*E. citriodora*) like bedding petunias; an avocado, which has been fruiting magnificently for years; three olive trees close together in the front garden because of my then prejudice about single 'specimen' trees; a brown plum pine (*Podocarpus elata*), now a very handsome tree, neighbouring an Irish Strawberry tree (*Arbutus unedo*), both with dark foliage. A principle I came to early in Perth is that the strong, clear light demands a strong statement from plants, and that dark cypress-green and the silvery grey-green of olives make a firm background, where the intermediate greens often look washed out.

Thirty-five years later, the trees are still there; so are the folk who bought the house. A few years ago, the house block was very prominent, like a single skyscraper in a flat plain – but the skyscraper was composed of tall trees rather than concrete. There is no doubt that I overplanted that block, but it was a good-hearted mistake, and now that trees are back in favour in Claremont, it doesn't stand out so much, although the next-door neighbour on one side still has a front 'garden' that is nothing but bare lawn, without so much as a shrub, let alone a tree. In a way, the contrast is quite effective, since the trees that we planted make such a dense and strong backdrop.

The second item worth recording is the form of the back garden. The back of the house faces north, sunny in winter. We put in a glass wall in the main room to the rear, opening onto a paved area with pergola and vines for summer shelter, and beyond that, a lawn of tough buffalo grass (*Stenotaphrum* species) flanked on three sides by bushes and trees, so the boundaries were hidden. I now call this form 'the clearing in the forest', thinking of its origins, and it is a form deeply embedded in the northern races of Europe. The clearing in the forest has a primitive appeal and a primitive history. Clearings, whether natural or artefact, were a comfort zone, and they gave you a clear view of an approaching enemy, whether man or beast. The precursor of the lawn was the grass cropped by domestic animals, often tethered for safety: goat, cow, horse. The dark, encircling forest formed the boundaries. Thus the three elements of this basic form are the open lawn, the closed, dark boundary, roughly semi-circular in form, and an unobstructed view to it across the grass – a view that has evolved from the view from the mouth of the cave to the picture window, but still giving primacy to one viewpoint, and hence guiding composition.

But if this form had dim ancestral appeal for me, it was not in fact typical of Perth in the sixties, nor of most of Australia, which used a form and style I have called Fletcherian, because it reaches an apogee at the Fletcher Jones headquarters outside Warrnambool in Victoria. The most obvious feature of the Fletcherian is its insistence on order. Neatness has been raised to a primary goal. Shrubs are pruned and clipped, lawns are immaculate, not a leaf out of place, there is a strong emphasis on flower colour, and a skilled succession of annual bedding plants, from Iceland poppies in winter through, say, nemesia and daffodils in spring, petunias in summer, asters and chrysanthemums in the autumn. Trees and shrubs are valued as horticultural specimens, and generally well tended by these standards. Unity of composition is subordinate to individual display, and foliage and form are subordinate to colour. The antecedents of this style are to be found in Britain, in the estates especially of the nouveaux riches and in municipal bedding displays, but also in the rapidly growing suburbs of the nineteenth century, especially in the industrial Midlands, where horticultural skills could be practised on very small housing lots, and the breeding of prize dahlias and chrysanthemums, or new cultivars of carnations, became matters of intense local pride and rivalry. The social origins of these changes are complex and obscure: they are in part a by-product of rapid industrialisation and urbanisation, and to some degree of that global colonial expansion which so increased horticultural choice, creating the global supermarket with the world's flora on the shelves. The obsession with order, both in Britain and even more so in Australia, was in my view the obverse of the fear of disorder in a world where the loss of respectability, from some minor misfortune or peccadillo, was irretrievable and terrible in its consequences. (This theme was introduced in Chapter 16. It continues to fascinate me, because understanding it must form a significant part of understanding our culture.)

I was in reaction to all this. I wrote *Swan River Landscapes* and *A Sense of Place* in the late sixties (see Chapter 11 for a discussion of their themes), and part of the ideas in these two books struggled for expression in the Claremont garden. I ripped out a plumbago hedge along the front boundary, which sloped down to the footpath level, and made an informal embankment with boulders of dense limestone caprock. I was in love with limestone; this was all before it became fashionable. I filled the pockets in the embankment with various prostrate thymes, prostrate rosemary, campanula, silene, thrift, pinks, all Mediterranean, all ordered by mail and eagerly awaited from Norgates near Trentham in Victoria. Such plants were long unavailable in Western Australia. I continued the Mediterranean theme with those three olives and the *Arbutus*. The blistering summer heat took out a few of the prostrate plants, but most of them settled in happily.

My 'sense of place' was incomplete; it never crossed my mind to plant *Templetonia retusa* or *Spyridium*, which were probably cleared to build the house, but I did plant a chenille honey myrtle, *Melaleuca hugelii*, which I had admired bursting into November bloom along the limestone escarpment of Kings Park, near the Brewery (it still does it). I experimented with melaleucas, and had quite a collection, including a bed of low, almost prostrate species, such as *M. violacea*, *M. cardiophylla* with its heart-shaped leaves, and *M. pulchella*, with exquisite corky, twisted branches, tiny leaves, and delicate mauve claw-flowers. And on the nature-strip (or virgin strip, as Elizabeth Jolley, our diagonally adjoining back fence neighbour used to call it) – on the nature-strip, I planted *M. nesophila* grown from seed that I had collected on the south coast in its home range. This is the earliest garden use of which I am aware, which is not to say that it is the earliest, but at least it was not on sale in the West at that time, although Lullfitz was a brave pioneer, and I bought the other melaleucas from him. Ralph Boddy at Eastern Park Nursery in Geelong may have had it by the late 1960s – he had certainly collected in the West. And it may well have been in common use in California. Almost all our other

167

beauties were, long before our nurseries carried much more than Geraldton wax. The contorted trunk and papery bark were the attraction for me. My tree was exposed to the south-westerly gales, and it became storm tossed as it grew, with a striking Gothic asymmetry, offending all the canons of order then prevailing. Yet by the time we left Perth, the tourist buses that included Victoria Avenue as part of their scenic route used to pause to admire my tree. It is still there, although a little the worse for wear from the loss of a limb that had to be amputated.

Paddington in Sydney

We moved from Perth to Sydney, where I took up the Chair of History and Philosophy of Science at the University of New South Wales. We bought a terrace house in Paddington immediately above the Scottish Hospital, which then had grounds that were a tangled wilderness of Port Jackson figs. The lie of the land sloped steeply down to the harbour at Rushcutters Bay, and we were about one-third of the way down from the ridge at Oxford Street, facing north, with winter sun, and protection from the cold southerly winds. Of course we were sitting on a platform of Hawkesbury Sandstone, like so much of Sydney, with meagre soil, but enough to grow a terrace garden: we painted the front door a smashing burnt orange, flanked the stone steps with wooden tubs, a cumquat tree in each. We paved most of the small front yard with flagstones, except for a large pocket near the wall parapet (topped with cast-iron railings, dark green). The pocket was planted up with a group of something or other, I think *Gordonia axillaris*. The back garden had a big camphor laurel at one side, a pergola right across the 6-metre width of the lot, with wisteria, bougainvillea, and a few other odds and ends. Most of the ground level was flagged. We were lucky to have a lane on one side and at the rear, with car access, but a terrace is a terrace, and the building form virtually dictates the garden.

Hawthorn in Melbourne

From Paddington, we went south. I became Director of the Centre for Environmental Studies at the University of Melbourne, and we lived in Hawthorn. Once again we were on a fairly steep slope, the northern slope to Gardiners Creek, which you encounter as Glenferrie Road swoops down from Scotch College to the creek, under the freeway, across the rail tracks, and up again to Toorak and Malvern Roads. We were two blocks east of Scotch. The front of the house faced south, and the people from whom we bought it had planted the steep slope to the street with eucalypts, hakeas and grevilleas. We left it alone – the fairly dense planting gave privacy from the street, it needed no supplementary water, and very little weeding either, since twigs and leaf litter inhibited most growth at ground level. The trees were a mixed bag, neither a natural association nor an arresting composition, but it worked. The back garden needed only a little effort to turn it into yet another 'clearing in the forest', a form to which I again turned instinctively. It was a fairly common form in Hawthorn-Kew-Camberwell-Malvern, much more so than in Perth, and much easier to establish and maintain, too. It was well-suited to suburban life, hence its popularity. The semi-circular backdrop gave enclosure and privacy; the lawn provided play space for children, and its openness still allowed you to see the enemy coming, especially if you installed security lights to deter would-be burglars at night. We never gave security a thought in the Claremont of the late sixties and early seventies, but by the eighties it was critical in Paddington and getting on to the agenda in Hawthorn.

That garden was pleasant, but no great creative energy went into it. All I remember vividly was my discovery of the American shadbush, *Amelanchier canadensis*, which I planted a few metres from the kitchen window, lovely at all times of the year; the new leaves were in silken sheaths, like pussy willow, the flowers were prolific, like small white plum blossoms, followed by handsome fruit which the birds liked. The leaves were a fine tracery, and the bare branches in winter had a Japanese delicacy.

Richmond in Melbourne

But I left it for Richmond, partly because it took an extra three-quarters of an hour just to get across the Yarra on the way to the University of Melbourne. Richmond was my most determined garden, nothing like anything else in the neighbourhood. Its two major characteristics were that it was full of 'choice' plants, and that there was no lawn, either fore or aft. The back garden was a kind of Dalmatian meadow, with a few native grasses thrown in for good measure. Up to a point, it reflects an élite Melbourne garden culture. The Melbourne of Chandlers Nursery, and Dicksonia Rare Plants, Stephen Ryan's luxury liner of pedigreed exotics at Macedon, Yamina Rare Plants at Monbulk, Bleak House Roses at Malmsbury, Lambley Perennials. I haunted them all, carrying back to my Richmond backyard all the jewels in the horticultural crown. Soon *Arbutus* X *andrachnoides* was in the ground, aspiring to grow its superb trunk of ox-blood red; I had all the clematis species Chandlers could provide, including the lovely little *Clematis nepalensis*, and *C. armandi*, with its glossy evergreen leaves and apple blossom flowers. I even had a white form of the Chilean bellflower, *Lapageria rosea*, and of course some species and near-species roses, such as *Rosa* Nevada, a *moyesi* hybrid. In general, however, one avoided hybrids and cultivars, and went for the true species, the idea being that the delicacy and natural form of plants, including the relation between flower and foliage, had been lost in the search for flower size and colour.

169

I am a little ashamed to look back on all this, but only a little. I learned a lot about plants. Many of these 'choice' plants were indeed aristocrats of the plant world, and most of them grew comfortably enough in Melbourne's temperate climate. In any case, rarity was not my only concern: I paid a great deal of attention to form, both of the plants and of the whole, and the result yielded a garden that was small, yet that invited moving inspection. It was no longer a picture from a preferred viewpoint, and there was 'always something happening'. The karst landscape was fairly successful. I found some very sensuous hard limestone stepping stones, rounded in plan but flat enough to walk on ('found' is a misnomer: I hauled them up a cliff, one by one, from a secluded beach near Aireys Inlet, doubtless illegally, but they would have been pulverised by the next winter's storm). These paths looped around the meadow, filled with low perennials, mostly Mediterranean, like *Artemisia schmidtiana* and *Euphorbia myrsinites*, *E. robbii*, Herb Robert, and so on: if you flick through Jennifer Wilkinson's book, *Herbs and Flowers of the Cottage Garden* (1989), you will find most of them, although this was no cottage garden, but a meadow.

It worked. It looked good and didn't require too much supplemental watering or fertilising. Snails and pests were not a serious problem, since most of these plants were adapted to the conditions under which I was growing them. Weeding was not too arduous, either, as there was little bare or disturbed ground. But boundary maintenance, a form of weeding, was a chore: the stronger growing plants were always ready to overgrow and swamp the weaker. And I spent much time picking up leaves. A meadow is tree-less, a nullarbor, but my neighbours had deciduous trees, and I had a few trees and leaf shedders too. A bed of leaves over these meadow plants would shade and rot them, and in the end, I was glad to leave the garden, rare plants, meadow, the lot. It was a brave try, and I do miss some of the proud beauties, but it was time to move on, back to the sun and the west.

Fremantle

So we began again at the other end of this dry, flat land – in Fremantle by the Indian Ocean, which we look at while we have breakfast. We eat it on the upstairs back verandah of an old, two-storey limestone house on a steep north-facing slope, built in the convict years for the chief warden of Fremantle Gaol. We overlook the Port of Fremantle, and beyond it, the turquoise sea, with Rottnest on the horizon. We like the port, with its giant gantries for the container ships, the wheat silos and silver array of big oil tanks – the dreaming cylinders of North Wharf – the liners as they come and go, and those extraordinary vessels we have come to think of as 'The Sheep Hilton', which light up like a luxury hotel at night, although the sheep will be far from clover on their long, crowded run through the torrid zone.

We pay for these pleasures. The house is sited by an intersection of two busy roads, and great lorries thunder past on their way to the port, or stop at the traffic lights with a squeal of air-brakes, but one gets used to it. We have known worse, and limestone walls nearly half a metre thick with double glazing on the most exposed windows exclude most of it inside the house. There are many rewards. The neighbour below is very quiet – Fred Samson, long-time mayor of Fremantle like his father before him, left his home to the state. Sir Frederick died in 1972, and the house is now a part of the Museum of Western Australia. The long stone wall of his laundry and carriage house is our back fence – designed by Sir John Talbot Hobbs in 1898. The intricate Victorian roof of Samson House and large old trees in its garden make the lower frame to our view. Another reward of living where we do is to be able to walk down hill to the town centre, two blocks away.

The house is not ornate, but strong and simple, with good proportions, and because the walls are so thick, the doors and windows have deep, splayed reveals. The windows all have glazing bars and panes, and this, with the depth of their setting, gives a beautiful light. But the house has been abused: it was tenanted by a government department, using it as offices, and it became run down. The 'garden' has been neglected for years, and part of the site had been made into a large car park (which had come to be seen as public, so we spent some time repelling invaders in our first few years).

This is my fifth garden, so the first steps are automatic, but the context has been different each time, and dramatically so this time. First, of course, we surveyed the site. A big olive, an apricot, a grapefruit, and a big old mulberry with delicious fruit. A sickly Moreton Bay fig, medium size. 'It's got to go', said my wife, so it went. Three more figs with a delicious black fruit. A superb 'Ilyarrie' or *Eucalyptus erythrocorys*, with its great warty fruits. A poplar, and too many suckers – since removed, as were two Chinese elms, which were poor specimens, and the Ilyarrie and a graceful little weeping mulberry, both of which were handsome but misplaced. There are two big coral trees (*Erythrina* cf. *indica*). These are so common and so easy to grow in Perth that they are very much taken for granted, but they are a good shade tree in summer, with large heart-shaped leaves, and their profusely born and brilliantly red blooms are winter cheer on bare branches, against a clear blue sky.

The soil, on the other hand, is not a joy, and it is a wonder that anything will grow in it at all. Yet plants will thrive if they are chosen well, sited carefully, and treated according to their needs. These are the universal rules of good gardening, but ones that must be rigorously applied here. The soil is not sandy, but sand, and not the deep, well-draining, rather coarse yellow sands of Claremont and Peppermint Grove, a few kilometres from the coast (the Karrakatta Sands), but a fine white calcareous sand, almost devoid of organic matter, and water repellent. Before planting, therefore, it is essential to treat the planting area with a surfactant – there are several commercial products, with names like 'Wetta-soil'. Without it, the water just runs off.

On our site, there is cap-rock just below the surface. This is a very fine, hard limestone, formed by solution and redeposition of the lime in the aeolianite (the old calcareous dune sands). It is a fine building stone, if you can find a stone-mason who knows how to work with it. Our house seems to have been built from stone quarried on site, by cutting into the hillside, and we are doing a little extension and repair with our own cap-rock. It makes planting hazardous when you encounter it half a metre or less below the surface, although some plants seem able to force their roots through crevices. To plant trees, I have to probe until I can find a 'solution pipe', the hard-walled cylinders that penetrate the cap-rock. Charles Darwin analysed their origins correctly when the *Beagle* put in at Albany. The root-tips of some trees secrete an acid which dissolves a slow way through the dense limestone, eventually to form a vertical shaft.

171

To the difficulties presented by a shallow, sterile soil, one must add further hazards: an exposed site, subject to very strong winds in late winter and spring, at times with a high salt content; and constant vehicle emissions. The rainfall is bounteous, but restricted to less than half the year; indeed, more than two-thirds of it falls in the five months of May to August, and there is no effective rainfall from November to March. Fremantle has a slightly lower annual rainfall than Perth, with 768 mm, which is nearly twice that of other west coast cities of similar latitude, such as Valparaiso in Chile, and Mogador on the west coast of Morocco – for which we have the relatively warm coastal waters off Perth in winter to thank. The rainfall is also fairly reliable, significantly more so than that of south-eastern Australia.

The bottom line for gardeners, however, is that the summer months are very dry. The choice therefore is between constant watering, which is a very common practice here, or drought-tolerant plants. We set out to design a garden that would flourish without constant watering once established; a deep soak once or twice a month is all we intended, and that was our choice at first. It immediately excluded lawn grass, and demanded paving in all areas of wear, as this sandy soil will not stand up to traffic.

We have not altogether put this into practice, for several reasons. We have certainly done a great deal of paving, and now have more bricks than most people have garden. We have also avoided all the water-addicts: no hydrangeas or azaleas or the plants from the humid tropics that are so popular here. But we do have a lawn, because we found that it was the only convenient way to maintain the slope that it now moulds rather gracefully, but that three years ago was a metre high in wild oats. Weed invasion is relentless here. There were also some design reasons, which I shall come to presently, for not living entirely within the natural rainfall. And of course we weakened – there are always a few plants that insist on finding a place, for example the roses. We have three stone buildings on the site, the two-storey house, a small shop, which we lease as professional offices, and a cottage, supposedly built around 1840, although the records are not good. The east wall of the cottage invited *Rosa laevigata*, 'Sparrieshoop', and the single white and deliciously perfumed variant of *Rosa banksiae*. They are all on fortuneana rootstock, which can cope with the sand and dry conditions, but they still need soaking at least once a week.

A final consideration is that there is a 120-year-old well by the house, 20 metres deep, cut all the way down through the limestone, and the original source of domestic water. We tested it, and found that the water is of good quality, although of course alkaline. There is only about half a metre of water in the bottom, but there is a good flow, and continuous pumping is feasible. So we reticulated about one-third of the garden. The rest is paved, or else planted with species that need no or little supplementary watering.

The choice of plants had to be made within a range defined by the ability to tolerate – or enjoy – heat, a degree of drought, and alkaline soils. The latter immediately rules out the majority of Australian native plants, including nearly all the banksias (*Banksia praemorsa* is one of the few exceptions, and we have one that is coming along beautifully). Just what thrives under these conditions is not always easy to tell, and the available books are no help at all, not even Lord and Willis (1982), splendid companion though that is. For example, some plants of the humid tropics are remarkably drought-tolerant. Frangipani is one example, but then it is a semi-succulent. The poinciana (*Delonix regia*) is more surprising – there is one near us in a neighbour's backyard that never gets watered, sitting on cap-rock, and it is a superb specimen.

So we prowled the neighbourhood to see what grows well and looks good, and also to see what 'works'; that is, what looks good against a limestone wall, what plants go well together, how different plants respond to wind exposure and so on. These are the horticultural and aesthetic constraints. Then there was the constraint of availability, which I will come to in a moment. But there was also a basic choice of style, which we pondered long and, for a time, indecisively.

Choosing a style

Fremantle has as many garden styles as any other Australian town, ranging from the utilitarian – rows of tomatoes and beans in the front garden – through to the latest, up-market 'cottage garden', with standard roses and box hedges, the like of which have never been seen here before. Rejecting the extremes, there are still three basic styles in Fremantle. They overlap, but they are nevertheless quite distinct in their feeling. One is Victorian; this is a solidly Victorian city, with many buildings from the 1890s, and a few that are older. They are nearly all either limestone, or cement render usually painted in pale or stone colours. The America's Cup sold a lot of paint here. Date palms, Norfolk Island pines, Moreton Bay figs, roses, buffalo-grass lawns, a few hardy shrubs, a few roses, a planting scheme that is basically sym-metrical, and a bed of annuals for colour. These gardens look pretty good; they suit the houses, and of course the trees and shrubs are mature and handsome. The roses do remarkably well and are very popular everywhere.

The Mediterranean garden overlaps the Victorian; add olives, figs, mulberries, lemons, *Arbutus unedo* (known in Australia as the Irish strawberry tree; it does extend to Ireland, but is typically found in the *maquis* of southern Europe), cork oaks, oleanders, rosemary, lavender, planes, Italian cypress. Many of these plants came by way of the Cape, whence came the freesias, which grow wild and flower freely all over the limestone hills. So, alas, did fennel and *Pennisetum alopecuroides*, although the latter has very attractive plumes in flower and we have used it at the front of a border, where it often provokes admiration from people who think they are seeing it for the first time, although it grows wild all over nearby Buckland Hill. Of course the limestone suits all these plants very well, and they can all survive the summer with little supplementary water.

The third possible 'style' is not a garden, but the natural environment itself. Fremantle is one of those relatively few towns in Australia which still relates to its natural environment, as do parts of Sydney on the North Shore. Its bones are good and they show in limestone outcrops, hills, cliffs, cuttings. Cantonment Hill and the grounds of John Curtin High School still carry *Melaleuca pubescens*, here known as the Rottnest Island tea-tree, with its dark umbrella, and the Swan River cypress, *Callitris preissii*, equally black-green, superb against the clear blue skies. *Melaleuca huegelii* flowers prolifically in early summer, with a profusion of creamy spikes. *Acacia rostellifera* is common, wind-sheared into a dense mound which protects the soil and moulds the landscape. Cockies Tongue (*Templetonia retusa*) puts out its brick-red pea flowers in late winter, and the Rottnest daisy (*Brunonia australis*) has sky-blue flowers in spring. The natural environment of dune plants also survives along the coast and was restored at Bathers Beach near Arthur Head, with Bicentennial funding. Alas, the money now seems to have run out, with no funds for the maintenance of the very good work that has been done. We have used all of these local plants and more; in fact this natural vegetation is so important and so fitting that our first thought was to restrict ourselves to it. A few recent settlers have done so, where new houses of limestone have been built on virgin sites in East Fremantle near the river, and the effect is stunning, an object lesson in relating to the natural environment. This is quite different, by the way, from the 'native gardens' which abound on new subdivisions, using an ill-assorted melange of plants which have in common only the fact that they are of Australian origin, which of itself guarantees neither harmony nor design, in fact not even survival.

173

An eclectic choice

But we have a house from the last century, and there is a good range of trees in our garden already, mostly what I have called Mediterranean. So we rejected the temptation to be purist and decided on a Mediterranean garden, but eclectic in the choice of plant materials, provided that they be drought- and lime-tolerant. So far, we have planted eight species of artemesia and four of lavender, lavender cotton, good old *Agapanthus*, utterly reliable here as in so much of Australia, a cork oak

(*Quercus suber*), the Italian evergreen oak (*Q. ilex*), the Californian liveoak (*Q. agrifolia*) and the Algerian Oak (*Q. canariensis*), which is rapturously happy, as is everything from the Canaries, here. There is a mass planting of Swan River cypress, Indian Hawthorn (*Rhaphiolepis umbellata*), and oleander.

It is frustrating that you often can't get handsome plants that you can see growing well around you in old gardens. There is a *Pavetta* species, probably *P. capensis*, with glossy leaves and heads of fragrant white four-petalled flowers in a garden near us. The nurseries haven't heard of it. There is a wonderful *Bauhinia hookeri* in the Crawley Bay reserve. It took me a year to identify it, and another year to acquire one. My *Gardenia thunbergii* came from Melbourne. (It likes the warmth, but not the alkaline soil, and I think it will have to go.) But even quite common things were hard to get five years ago, although availability has since improved; for instance the single-flowered oleanders. The double flowers, which are readily available, go a dirty brown and hang on the bush before they fall, whereas the singles flower and then fall at once, without a long drawn-out deathbed agony. The only varieties commonly available a few years ago were a harsh lipstick pink double, and a dirty white double. We found a stock of single pale pinks and bought them all. The single apricot was hard to find. The remedy, of course, is simple. Propagate your own. So we now have some apricot oleanders coming along nicely, and a primrose yellow, and a champagne-coloured one and one that is almost white with a blush-pink throat, all blooming profusely. The local cemetery proved a good hunting ground for cuttings, which will grow roots in a bottle of water in a couple of weeks. Various seeds we brought from Melbourne. *Lilium filippinensis* did so well for us in Richmond, it was worth a try here, where it does even better, and a few bulbs of *Amaryllis belladonna* 'Hathor' are now clumping well. All the Cape bulbs thrive.

We are happy with our choices. We began planting immediately after Christmas in 1990 and all but one plant (the one that was replaced) have survived four searing summers. The odd thing is that our choice is so far out of the current fashion, which is either for 'native' gardens, mentioned above, or a tropical luxuriance of *Hibiscus*, palms, *Bougainvillea*, *Acalypha*, *Schefflera* (the Queensland umbrella tree, it is called here), *Ficus benjamina* and the like. Such gardens can look very lush if enough water is poured on, but there is an obvious sense in which they are against nature. Turn off the sprinklers for a summer and they would wither and die. That there is something not quite right shows in the colours. This is not the light of the humid tropics, grey, subdued, in which bright colours glow like jewels. The light here is hard, clear, bright – and it needs the black greens of cypress, whether native or exotic, and the greys of lavender and olive, or *Calocephalus brownii* and the coastal *Olearia* and *Rhagodia*. The lush look is best kept to a courtyard, excised from the broad environment. Out in the open, it looks unnatural, and the colours rather garish – or so we think.

Structural decisions

The hardest decisions to make and put into practice were structural – the difficulties being that half the area was a construction site for the first two years, while major repairs were carried out on the buildings, with debris everywhere; and that the structural requirements were costly, and had to wait; and because design decisions had to be made. The sloping site, taken with the orthogonal shapes of and between

174

The passion for neatness; a typical suburban front yard in Perth.
GEORGE SEDDON

Our brick garden in Fremantle. GEORGE SEDDON

The Royal Crescent at Bath: a response in south-west
England to the Arcadian taste and the pastoral
economy of the day, one which favoured pasture and
clumps of trees for shade and shelter. GEORGE SEDDON

Trees in a clump, with a shapely canopy and a grassy
floor. *Melaleuca lanceolata* creates a version of the
Arcadian at Phillip Island, Victoria; not like Bath, but
representing a taste that has the same origins.
GEORGE SEDDON

A National Parks Ranger in Victoria with rabbits, cat and fox;
the cat is as big as the fox. DAVID INGRAM

Some Australian landscapes are of great age. The Laidlaw Range in the Canning Basin, WA, was a Devonian calcareous reef some 360 million years ago, and the small, isolated rocky outcrop in the middle distance (left) was an atoll—just add water. The old reef was buried by later sedimentation, uplifted, and then revealed again by erosion.
PHILLIP PLAYFORD

The arid zone supports the greatest diversity of lizards in the world. A 'dragon' or agamid lizard (*Pogona vitticeps*): an insectivore with a high body temperature when active, and good thermo-regulation. To warm up in the sun after a cold night the dragon flattens its body to almost double its breadth; to keep cool in the heat of the day it rests vertically on the trunk of a tree out of the sun.
JOHN HANRAHAN

Termite mounds in south Kakadu National Park. Termites are the primary reducers of cellulose in much of Australia, especially the north: 'They are not the inert tombstones they appear to be' – they are stationary cows. GEORGE SEDDON

Fire in the mallee. The trees are fire-adapted, and will usually sprout anew from lignotubers below ground. Stock, fences and farmhouses lack this useful adaptation. GEORGE SEDDON

the three buildings, all linearly aligned with the boundaries, seemed to dictate formality, so we built a series of descending terraces with robust rectilinear retaining walls of limestone. Only the eastern hillslope, still a more or less natural slope, was left to run free. The old quarry east of the house from which much of the building stone had been taken when the house was built in the 1870s was excavated further to make a level floor that drained away from the house, and this was paved with old brick, the two intersecting axes now marked with a fountain (see colour illustration following p. 174). Our stone-mason hails from Carpentras in the Vaucluse, north-central Provence, and he built us beautiful walls that will endure long beyond our life span. The result is a series of geometrically defined spaces, with three intimate courtyards and a large one on the northern boundary, each with different use characteristics. They are linked by axial walkways, all paved with brick. The forms are thus what is sometimes called 'architectural' rather than informal and 'naturalistic', although these are not very helpful terms. The garden is 'designed with nature' in that the plants chosen are well-adapted to their setting, even though they are not all indigenous, nor set out in irregular, sinuous forms.

This is not a 'clearing in the forest', either, and it is certainly not a meadow. It is nearer to another primary garden archetype, the oasis. There are three water features; the fountain, which gives the gentle plash of falling water, and two small ponds in natural cavities in the dense limestone caprock. One is planted with nardoo (*Marsilea* species), an aquatic fern whose sporocarps were ground for food by Aborigines; the other with water lilies. Paradise fish and fan-tails keep the water clear and take care of mosquito larvae. Water gives a different kind of focus to a garden, yet this one is also outward looking, and the boundaries are pervious, largely because of the slope and difference in levels.

175

We enjoy this garden, which is just as well – I shall not make another. There is more maintenance than I intended, but there always is. We spent the first few years nursing plants through the fierce summers to keep them alive. Now they are established, we will spend the rest of our lives hacking them back to keep them within bounds. Plants that do well here do too well; *Romneya coulteri, Erigeron karvinskianus, Lilium filippensis, Geranium sanguineum* are among our worst weeds, while *the* worst are *Lobularia maritima* and purple toadflax, *Linum purpurea*. But we count our blessings: no frosts, and a garden the year round, much to be preferred to that grey, sodden mess that is all that is left of the garden for six months of the year in some places we know. Most important of all, of this as of all gardens, it is always possible to make changes for the better, or at least to dream about doing so. It is one area of our lives over which we have control.

References

Lord, Ernest E., and Willis, J. H., 1982, *Trees and Shrubs for Australian Gardens*, Port Melbourne: Lothian.

Seddon, G., 1995, *Swan Song: Reflections on Perth and Western Australia, 1956–1995*, Centre for Studies in Australian Literature, University of Western Australia.

Wilkinson, Jennifer, 1989, *Herbs and Flowers of the Cottage Garden*, Melbourne: Inkata Press.

19 The garden as Paradise*

Decisions and choices about gardens are not made in a vacuum. They are culturally mediated. A multitude of forces is at work, including, in our own day, the influence of fashion, travel, reading – sometimes, perhaps, even the influence of conferences. At a deeper level there are also some concepts with a long history. One set of concepts is expressed in Western cultures through a group of related words. The words are: Eden, Paradise, Arcadia, Utopia.

The relations between these words are complex and shifting. Most dictionaries, for example, give Eden and Paradise as synonymous, yet we all know that they are not interchangeable. They have a different etymology, different history, and a different range of application, even though the differences consist in subtle nuances. It seems appropriate to explore these differences a little, although necessarily in a superficial way, since they range across more than two thousand years of cultural history.

Utopia

Utopia is a good word to begin with because its origin is known with precision, and because it makes explicit a characteristic that in the others is usually implicit only. Utopia is a critique of the actual. The word was coined by Sir Thomas More in 1516 as the title for a book, conjoining two Greek words, où = not; topos = place. Utopia is an imaginary island, depicted as enjoying a perfect social, legal and political system. Sir Thomas More's England shared with Utopia the fact of being an island, but the resemblance stopped there. He does not present his imaginary island as a practicable alternative, but as a way of highlighting the imperfections of the world around him and of strengthening the will to improve them. This form of social critique is widespread, and pops up in some unexpected places. It is widely claimed today, for example, that Margaret Mead's well-known anthropological works on Samoa and the Trobriand Islanders are much more a criticism of the narrowly

*First published as 'The garden as Paradise', *Australian Garden History: Journal of the Australian Garden History Society*, vol. 8, no. 4, 1997, pp. 8–12.

puritanical United States in which she was brought up than they are about the real-life behaviour of the people she purported to be studying.

As noted above, the emphasis of Sir Thomas More's Utopia is on the social, legal and political system, as one might expect of the chancellor of the realm, but he nevertheless goes into some detail about the gardens (and makes them sound remarkably like Australian suburbia, a point that its critics might consider).

> The stretes be twentie foote brode. On the backe side of the houses through the whole length of the streete, lye large gardens inclosed round aboute wyth the backe part of the streetes. Every house hathe two doores, one into the streete, and a posterne doore on the backsyde into the garden . . . They set great store by their gardeines. In them they have vineyardes, all maner of fruite, herbes, and flowres, so pleasaunt, so well furnished, and so fynely kepte, that I never sawe thynge more frutefull, nor better trimmed in anye place. Their studie and deligence herin commeth not onely of pleasure, but also of a certen strife and contention that is between strete and strete, concerning the trimming, husbanding, and furnishing of ther gardens: everye man for his owne parte. And verilye you shall not lightelye finde in all the citie anye thinge, that is more commodious, eyther for the profite of the Citizens, or for pleasure. And therefore it maye seme that the first founder of the citie mynded nothing so much as these gardens. (More, 1516, p. 90)

Sir Thomas More's word 'Utopia' has had shifts in meaning since 1516: one that is used in the planning literature is to make it the antonym of 'dystopia', the bad or dysfunctional place, compared with 'the good place', although strict etymology would require that to be spelt 'Eutopia' ('eu' meaning 'good' in Greek, of which More, a good classicist, was well aware, his title intending the two possible meanings). A much more familiar shift in meaning is that a 'Utopian scheme' is now seen as impractical, the idle dream of the fanciful who know nothing about the real world, and in this sense the term is dismissive, even contemptuous. This shift, however, emphasises one further characteristic of all of the words with which I began: Utopia is *unattainable*. Attitudes to the unattainable then diverge. The unattainable can be seen either as a delusory will-o'-the wisp, leading us astray, making us long for something we can't have – we would be better off without it, getting on with the realistically attainable. Or it can be seen as the energising dream that leads us on creatively to achieve far more than we could ever believe possible without it, even though the perfection always remains just beyond our grasp. This tension underlies a great deal of writing and thinking about the design and practice of gardening (and much else).

From Sir Thomas More's 'Utopia', I draw three points: that there can be no ideal world without gardens, and even, perhaps, Best Kept Street and Tidy Town competitions; that all Utopias are a critique of our mundane surroundings; that they set up unattainable goals – which can be seen either as inspiration or distracting folly (and I might add that every gardener I know is deeply ambivalent about these two attitudes, sharing now one, now the other).

Eden and Paradise

'Eden' and 'Paradise' have a longer history, and they are not so much imaginary as mythical, although they can be understood quite literally by fundamentalist Christians and Muslims. Eden shares with Utopia the characteristic of being unattainable: it is a lost world, from which we have for ever been shut out by the original sin of Adam and Eve. Paradise, in one of its uses, is an attainable reward, to those who are literal in their faith, for living virtuously. Curiously, however, 'Eden' seems to have a much more concrete reality than 'Paradise'. It has been notionally located as an actual site somewhere on the Anatolian Plateau, for example, not without reason, given that most of our common fruits had their origin there. One recurring feature of imagined Edens is that there was always plenty to eat without much effort, and a range of fruit trees with an extended fruiting season covering much of the year seems to fill this need well. But it has also been used metaphorically of new-found lands by Europeans, if they were seen as being fertile, without climatic extremes, and, as a rule, inhabited by men and women perceived as living with noble simplicity.

Adam and Eve were naked, and Eden has always suggested an innocent sexuality – an element of Margaret Mead's Samoa of the Noble Savage, who was earlier identified in the European imagination with the American Indian – not, alas, in European behaviour, the identification being restricted to the literary imagination of the salons. This search for primitive nobility and innocent sexuality in the naked or semi-naked savage is not restricted to the Christian myth of Eden: it is fully developed in Tacitus, for example, expressed in his *Germania,* in which the forest dwellers to the north of the Alps were seen as having all the primitive virtues that had once supposedly characterised the Romans, but had been lost in a society that was urban, corrupt, autocratic, licentious, luxurious, effete (see Schama, 1995). This is another key theme, and it applies in varying degrees to all of my four words; the 'garden', however conceptualised, is antithetic to the vices of urban blight.

In Paradise, the sexuality was not always quite so innocent as it is portrayed in, for example, the almost puritanically innocent Adam and Eve of the sculptures and paintings of mediaeval Germany. Paradise might offer a more abundant sensuality, more like the guiltless but very active sexuality portrayed by Margaret Mead. The Muslim Paradise was (is?) conveniently furnished with houris, voluptuously beautiful nymphs with eyes like gazelles (that is the Arabic and Persian derivation of the word). This, it is apparent, is a male-oriented concept of the delights of Paradise: sexual equality would presumably require that there be a plentiful supply of graceful and virile young men at hand to pleasure the ladies who had made it to Paradise, but the Muslim concept is reticent on this point. The idea of a rich sensuality inevitably degenerates into license and sexual ambiguity in some versions of this theme. The Forest of Arden was the site of imagined indulgences of many kinds, including a loosening of the gender roles ('If you go down to the woods today, you're in for a Big Surprise').

The loosening of defined roles relates not only to gender, but to social hierarchy, another recurrent theme that has been expressed in a variety of ways through time. 'When Adam delved and Eve span, Who was then the gentleman?' is an assertion of the rights of man, but the democratic ideal was located by Tacitus in the Hercynian forests of Germany; Shakespeare inverts the social order – for a time – in the Forest of Arden, and Robin Hood in the greenwood, both part of a long tradition. The 'garden' in its various manifestations is usually (but not always) seen as a place in which distinctions of rank are laid aside. This carries through to the present in very minor but recognisable forms. There is, for example, a freemasonry of gardeners. 'Real gardeners' recognise each other and accept each other at once as equals, whether they be Dame Elisabeth Murdoch, Lady Law-Smith, or Jack's Jill. The Open Garden Scheme in Australia could not work without this recognition that gardens are for gardeners, and thus for sharing among equals.

Of course the dictionaries are correct in listing 'Eden' and 'Paradise' as synonyms, and the different nuances that attach to them are never used consistently. As well as differing in tense, however, they differ etymologically. 'Eden' has a Hebraic origin, associated with the word meaning 'delight' – hence the garden of earthly delights. Whether as part of the biblical story, Christian and Jewish, or as an ancestral memory of favoured valleys in Asia Minor, 'Eden' is earthly. The *OED* gives its primary and secondary meanings as '1. The first abode of Adam and Eve, Paradise, and 2. *transf.* and *fig.* A delightful abode, a paradise'. Both end up as 'Paradise', but Paradise has a much longer definition, and range of applications. Its origin is also Middle Eastern, but in this case, Avestic (Old Persian), with the original meaning of a large park. Its first recorded use in Greek is by the historian Xenophon, who used it to describe the parks of Persian kings and nobles. These were usually hunting parks, so they were extensive, and they contained wild animals, and this persisted as one of the later meanings as 'an oriental pleasure ground, *esp.* one enclosing wild beasts for the chase. *b.* Hence an English park in which foreign animals are kept. 1613'. When it is used synonymously with 'Eden' it refers to the *earthly* Paradise, whereas the *heavenly* Paradise is the abode of the blessed, either Christian, Jewish or Muslim.

The difference in origins is reflected in the metaphorical application to gardens: the walled garden, full of flowers, fragrance and sweet ease is an Eden derivative, whereas the English estates improved by Capability Brown and Humphrey Repton in the eighteenth century are paradisiac, offspring of the Persian hunting park, although the wild animals were usually restricted to deer, which were self-propelled lawn-mowers. These estates in England were the product of a pastoral economy: they are based on grazing rather than digging and delving – agriculture – and thus can also be seen as reflecting through the ages to our own day as two approaches to 'the garden' (using the word in the broad sense). One is actively interventionist, turning the sod, pruning, weeding, maintaining; a branch of agriculture. The garden space is enclosed, and animals, other than a dog or cat, are excluded. Animals are inimical to intensive agriculture. The other is more extensive and more expansive, less obviously bounded, relying for its pleasures on mature trees, groves of them where possible, grassy clearings (which we now call 'lawns') with an understorey of shrubs in the woods and at their margins, but no defined planting beds: in short, a pastoral system rather than an agricultural one, in which it grades imperceptibly into the 'Arcadian', my fourth word, and still to come. The persistence of lawns into contemporary suburban gardens in Australia is a very clear indication of the extent to which garden design is mediated through cultural history. Australia was settled at a time when the dominant social group in Britain was nurtured by a pastoral economy. The lawn is an Arcadian remnant.

This distinction between Eden and Paradise is neither clear-cut nor consistent. The nuances shift and change. The Eden of the Book of Genesis was not agricultural (the delving followed the Fall), but it was certainly not a hunting economy, and graphic representations show the lion lying down with the lamb. The dietary regime

179

at the Hotel Eden is not specified in detail: Eve certainly did not spend the day over a hot stove, so elaborate food preparation is out. One gets the impression that wild honey, milk (possibly) and lots of fruit, apples excepted, were the staples. In other words, close to the diet of the frugiverous primates of the Great Apes to this day. Is Eden a memory? It is possible to see the Fall as the beginning of agriculture, with its life of backbreaking toil, not so much a milestone of evolutionary progress but a consequence of increasing density of population. Anthropologists have used the phrase 'primitive affluence' of hunter-gatherer societies, and it has been argued that the Aborigines in Australia in the eighteenth century had a more nutritious and more varied diet, acquired with substantially less expenditure of effort, than their European counterparts. Perhaps we can see the gatherers in Eden and the hunters and herders in the Paradise myth, representing two strains of a pre-agrarian society.

Nevertheless, the Eden derivatives among our gardens do require intensive labour, and this is precisely because of the Fall. We can recreate imagined Edens, gardens of earthly delights, but their cost is unending hard work – pleasurable, perhaps, but work.

Arcadia

Arcadia or Arcady overlaps considerably in meaning with 'Paradise', but it is a product of the Greek and Roman cultures, not the Middle Eastern. It was supposedly drawn from a real place in the Peleponnese, one favoured by nature. In both cultures, it was imagined as a wooded rocky place, the haunt of satyrs, the realm of Pan with his sweetly haunting pipes. It is the myth of a pastoral economy, idealised by city dwellers, and to that extent, it can be read as escapist fantasy. The Theocritean poetry of shepherds, shepherdesses and their innocently amorous pursuits were in part a product of urban overcrowding, first in Greek Alexandria, later in Rome. It had a febrile revival at the French court, where there were eager swains aplenty ready and willing to pursue their reluctant loves through the woods. In England, an Arcadian-Paradisiac ideal became the basis of landscape design, as we have seen. Because of the Enclosures and the switch from agriculture to a pastoral economy on the large estates in the eighteenth century, a productive grazing estate could also be conceived as a *landscape* of shady groves and open meadows, and the 'garden' or park was a semi-natural, productive world of Nature refined of its coarser attributes. An apparently natural harmony between people and their setting is central to the Arcadian ideal, and it can be illustrated well in many of the older rural estates in the better-watered and more fertile parts of Australia, which were receptive to the Arcadian imagination, since they are all part of the pastoral economy (see colour illustrations following p. 174). The influence of the Arcadian dream is strikingly apparent in Australia from the self-image that so many graziers evince through the names they gave their land: the key word is 'Park'; Sefton Park, Camberley Park, Alton Park, Camden Park. How many 'Parks' do you know in your neighbourhood? They are not, of course, parks – they are sheep farms or cattle farms. A park is devoted to leisure pursuits, not to earning a living. It is only by invoking the Arcadian dream that the two can be yoked satisfactorily.

A good example of an Arcadian landscape 'on the ground' is the driveway and broad setting of Dame Elisabeth Murdoch's Cruden Farm at Langwarrin, on the outskirts of Melbourne, and one could use three of the words we have been discussing to name its parts. There is a walled garden that is a 'garden of earthly delights', which is to say, 'Eden', sheltered, bound, fragrant, full of flowers – but because of that apple-eating, requiring a great deal of hard work. Around the house there is a more open garden with grassy walks, shrubberies and many fine trees, more Paradise than Eden; this runs out to a working farm, with a gentle transition. But the 'farm', which is pasture and shade trees, is carefully groomed. The curving driveway itself is of gravel, not bitumen, which maintains the rustic feeling, but increases the maintenance. Beside the avenue of trees, there is a split-rail fence, and beyond that, the blond pastures of summer. Intensely Australian, naturalistic, but the effects are highly contrived, without any appearance of contrivance.

The critical use of terms

If we define each of my four terms carefully, and use them precisely and consistently, can they then have a more rigorous employment in the analysis of garden history? The answer is, alas, no. They will always shift and change. 'Eden', 'Paradise' and 'Arcadia' will continue to be used interchangeably on many occasions, and all are used with a Utopian context, in the sense of being ideals. The point of considering them at all is to gain insight into the complexities of cultural history.

Paradise translated

The European imagination has often located Paradise in the New World. The French have been specially fertile – Jean-Jacques Rousseau and the Noble Savage, Gauguin and his languorous Paradise in the South Seas, the 'Paysages Exotiques' series of the Parisian painter, Henri Rousseau, the customs officer who had never travelled, but worked from the Jardin des Plantes, the Botanic Gardens in Paris. His paintings pick up several of the themes noted already. It seems always to be spring. The colours are bright and clear. Apes play happily with golden fruits. Another French fantasy is the account of the voyage to La Terre Australe by Jacques Sadeur (1693), with which we began (see Prelude). Remember that there are only four large animals, and none of them dangerous, there are no venomous serpents, there are no troublesome insects and, happily, there are no flies. We live on fruit that ripens the year round – and so on.

181

The English were more pragmatic: to describe a new land as 'like a garden' or 'a second Eden' was generally a prelude to occupation. 'The country was a Paradise', said Thomas Arnold, of New Zealand. One early 1800s English traveller to Australia was impressed enough by what he saw being achieved here in New South Wales 'even under the worst auspices, and in a country filled with the dregs of our own . . . How much more then, must New Zealand flourish, which is itself a beautiful garden and capable of being rendered the most delightful spot on earth.' (in Park, 1995, pp. 323–4)

The sting is in the tail. If the place is already a beautiful garden, where the soil is so fertile that it provides readily for all the necessities of life, then what further 'rendering' is required? The answer is that part of taking possession is to put your own stamp on the land. Very often this has meant wiping out the qualities that made the place so attractive in the first place, and trying to reproduce the landscapes and gardens you left behind. Thus the garden in the New Worlds, the Americas,

Australia, New Zealand, can be seen as one of the tools of imperial power; it remains one of its symbols and the slowest to loosen its hold.

Future directions

John of Gaunt called his England 'this other Eden, demi-Paradise', a sentiment of the kind most often heard in times of war, when patriotic fervour runs strong. But it also expresses a mood of acceptance and celebration. Eden is here. Embrace the actual – or, in language less elegant than Shakespeare's, but still pungent: 'This is as good as it gets', says Paul Keating.

This mood is strong in Australia today: we should abandon the delusive dreams – and there is more than one. The Utopian dream of Gabriel de Foigny and his imagined voyager Jacques Sadeur gives way to the reality: there *are* flies, venomous serpents, destructive insects of all kinds. The English imperial dream is also dispensable, the dream of conquest and transformation. We do not need to achieve social standing by recreating Knole or Bodnant in hot and dry Australia. Macedon, Bowral, the Adelaide Hills, and Toowoomba are not the only places in Australia where it is possible to have fine gardens, although the gardening literature would lead us to think so. They are regularly described as having climates 'favourable to gardening', which means, of course, slightly more amenable to pursuing generally inappropriate goals than the rest of the country.

Nevertheless, the ambiguities remain. No matter how much we dedicate ourselves to ecologically responsible gardening, gardens do not look after themselves, and we cannot leave it to nature. The better introduced plants are adapted to local conditions the more likely they are to become dangerous garden escapes. The Australian environment everywhere is already a disturbed environment. Weed invasion is unending. If we aim for a garden that is productive as well as decorative, problems multiply. Where I live, in Fremantle, the olive and the mulberry look after themselves, and fruit generously. Lemons need some fertiliser and summer water, but are pretty tough. Figs, however, need constant baiting. They fruit prolifically, but are full of fruit fly. And so it goes. Moreover we demand a higher level of comfort and design in our immediate environment than the natural environment generally offers.

If the European imagination has often idealised the lands of the sunny south as a place of warmth and leisure where the flowers always bloom and the ripe fruits drop effortlessly into your lap, those who live and garden there know a different world – one where plants grow rapidly, but also senesce and decay rapidly, where weeds also grow rapidly, rampantly, where there are no winters cold enough to kill off predatory insects, where air that is both humid and warm encourages the growth of fungus, mould, scale, virus, black spot . . . , where the organic content of the soil oxidises rapidly and where heavy rain leaches the soil of nutrients, yet also where the heat of the sun is such that even a few days without rain can constitute a drought, and four or five months without it, a significant challenge.

We can and should redefine our concept of Eden, Paradise, Arcadia, using local idioms and in ways that reflect more of the actual. But there are dreams that are enabling as well as dreams that are disabling. The latter feed on hope of some celestial bonanza, unearned, and they are escapist. We can have a better world, but it will be here, not over the hill or beyond the blue horizon, and we have to make it with our own two hands. This is the sustaining dream.

Postscript

It is not usual for a book on understanding and conserving landscape to devote a whole section to gardens. Yet Voltaire thought it necessary to cultivate one's garden. My reasons are his, and others that were not his. One is that gardeners are, in fact, one of the most important groups of land managers in this country, since between us we manage more than 50 per cent of all urban land in Australia, that is, the land that carries 80 per cent of the population: land that is not vast in area compared with that managed by farmers, pastoralists, miners and state agencies, but greater in value and in resource consumption than all of them. *Gardeners are key land managers.* Our choices therefore lie not in whether but in how we manage the land. We would all agree that we must do it in an ecologically responsible way.

But most Australians are *not* gardening in an ecologically responsible way at present, although some of us are trying. If we take the sum of gardening in Australia, it is undeniable that the resource input is gross. As with the rest of the world, we are an extreme reflection of the consumer society. The consumption of land alone is impressive. Perth, for example, stretches north–south for 120 kilometres and east–west for 70 kilometres, and this huge area houses 1.2 million people, at a density lower than that even of Los Angeles, and the reason is that nearly everyone has their own garden. Moreover, this is not just any land – although the soils are of low fertility, it lies in the only corner of the state that has a temperate climate and a substantial and reliable rainfall. The other state capitals, which house most of our population, are also favourably situated.

Next, the water consumption is heavy: it averages 515 kilolitres per household, and it is estimated that 40 per cent of this goes on parks and gardens. Lawn grass is a major irrigated crop in Western Australia. There are now concerns about both the quality of the water and the rate of replenishment.

Moreover, water moves nutrients in solution. Since our parks and gardens are heavily fertilised with chemical fertilisers brought in from distant sources, there is an oversupply of nutrients to the coastal lakes and rivers, and this is causing eutrophication and algal blooms. There is a long list of resources consumed: topsoil is brought in and peat is quarried, along with other soil-amending agents. Nearly all of them are in quite short supply in the natural environment. Oil and petrol are consumed by garden machinery, especially by lawn-mowers. Large quantities of garden waste are generated, through weeding and pruning. This may be chipped and composted, but much of it is carted off to sanitary landfill sites. Waste disposal is an increasingly expensive component of our urban economy. Finally, a whole range of insecticides and fungicides is sold and used, and some of these toxic chemicals find their way into the food chain, especially for birds. It is difficult to get good statistics for these inputs. The Australian Bureau of Statistics (ABS) reports that Australians spent $910 million in 1991 on nursery and horticultural products. This equals $172 per household, but the ABS statistics cover sales from nurseries only. Supermarkets and chain stores would double that figure comfortably, and for Perth, the figure is likely to be much higher than the national average, given the more demanding gardening environment. Clearly we need to lift our game. How?

The *appropriate* goals, it seems to me, are fairly easy to spell out. To reach them is harder, and requires commitment. We could return to the gardening practices of our parents, although in their case, they followed from necessity, while in ours, we have been undone by availability. No more bags of peat moss, nor loads of pirated topsoil, nor chemical fertilisers, nor toxic insecticides. Stick to pyrethrum, garlic sprays, hose off the aphids, make your own compost and mulch, keep seeds, swap cuttings, and go easy on the water. Above all, learn what works in your neighbourhood. There has been a great and healthy interest in garden design over the last few years, and this is both desirable and overdue, but the search for an Australian *style* is a chimera. John Brookes, the English garden designer, urged recently that we should be incorporating motifs from Aboriginal culture to make our gardens more distinctively Australian. My response is that we have taken enough from the Aborigines already without trying also to appropriate their culture, and that, in any case, the goal is wrong. Good design is good design. We do not need an Australian garden design style. What we do need is better design, and to evolve sound Australian gardening practices, which must be frugal, and, of necessity, regional, even local.

References

More, Sir Thomas, (1516) 1910, *Utopia*, trans. Ralph Robinson, 1551, 1556, (ed.) George Sampson, 1910, London: G. Bell and Sons.

Park, Geoff, 1995, *Nga Uruora: The Groves of Life*, Wellington: Victoria University Press.

Sadeur, Jacques (Foigny, Gabriel de), 1693, *A New Discovery of Terra Incognita Australis, or the Southern World, by James Sadeur a Frenchman*, London: Charles Hern.

Schama, Simon, 1995, *Landscape and Memory*, London: Harper Collins.

THEME

V

Analysing: ideologies and attitudes

To write of conserving landscapes is to walk a tightrope in a high wind. Some of the gusts have been recorded earlier: for example, we see ourselves as operating on and external to Nature; we see ourselves as part of Nature. Our very being impinges on the environment at every turn. 'Take nothing but photographs, leave nothing but footprints', says the Sierra Club; sound counsel to backpackers up from Los Angeles for their wilderness experience – but their film consumes some of the earth's less abundant minerals; their footprints may scar alpine plants with a very slow recovery rate. In Australia, where most plants are adapted to low-nutrient soils, even a modest infusion of human wastes may kill, while the water in most of the lakes of Kosciusko National Park is  not safe to drink. It would require an extraordinarily conti- nent or costive backpacker to leave only footprints.

Thus one can peer down on one side of the tightrope at the Hands Off Everything Society, for whom conserva- tion means preservation and heritage means stopping the clock at an arbitrary point in a fictive past, largely a nostalgic invention. This is a mood I share from time to time, that the future's not what it used to be. Lacking confidence in our capacity to adapt and change creatively, the past seems more comfortable. Meanwhile on the other side of the tightrope, one can see the upturned faces of derisive economists who argue, with some justification, that resources are created by demand. The known reserves of all the major resources are greater now than they were when the Club of Rome first made its bleak predictions. It is a gross oversimplification to use the metaphor of the world as a kind of supermarket whose shelves are stocked with goods in two categories, one marked 'renewable', to be replenished at regular intervals, the other marked 'non-renewable', once used, gone forever. The economists have a point: yet, like most people I know, I am still vulnerable to waves of sympathy for the bleaker view. Rose-coloured spectacles fog up so easily.

The two chapters in this section are both early essays. Chapter 20, The rhetoric and ethics of the environmental protest movement, was written in 1972, about a debate that still generates more heat than light. Chapter 21, The perfectibility of Nature, is a 1974 review of a sweetly reasoned book on the same subject by John Passmore, who was at the time a Professor of Philosophy at the Australian National University in Canberra. Would that there were more like him. Both essays could have gone under Theme I, Talking: the language of landscape, or Theme II, Perceiving: the eyes and the mind, but I prefer them at this point for two reasons. One is that I want to keep that analytical sense going throughout the book, so it is useful to give it another airing. The second is that I was teaching History and Philosophy of Science at the time that these essays were written; it would be pretentious to call them 'philosophy', but they do have the flavour of conceptual analysis, and thus go beyond the two earlier themes.

20 The rhetoric and ethics of the environmental protest movement*

The environmental protest movement has occasioned much cheap rhetoric, and I propose here to look at some of its characteristic modes. In doing so I must first clear myself of potential misunderstanding. I do not for a moment intend to suggest that there are not grave environmental problems ahead of us. There are. Nor do I want to suggest that we should somehow disinfect all discussion about them from emotion. We can't. One sometimes hears senior scientists in universities and government employ say that 'we must get rid of all this emotionalism'. People will continue to feel strongly about matters that concern them so deeply, and such scientists merely show themselves unfit to comment on the human environment if they so patently misunderstand people.

Fear, guilt, self-righteousness and hate are the emotions most commonly exploited by cheap rhetoric, although there are others, from greed, despair and self-disgust to the noblest idealism, which is not, however, given an effective outlet in action, and is thus frustrated. The rhetoric of the environmental movement has at least three distinctive modes, all with a long ancestry. For ease of reference I shall call them, then, the Jacquard, the Revivalist and the Bucolic.

The Jacquards

The Jacquards are the most interesting of the three, although the Revivalists are by far the more common. They are named for Jacques, from Shakespeare's *As You Like It*, and they are the jakesmen, poised for ever above a cesspool. The melancholics are especially well defined in late Elizabethan and Jacobean drama, and their defining characteristic is disgust – a generalised disgust, rather than specific disgust with specific abuses, which is healthy and may lead to corrective action. The times themselves are out of joint: 'Fie, 'tis an unweeded garden'. The early seventeenth century and the latter half of the twentieth seem to have been the great ages of melancholics. They are intelligent, hypersensitive, civilised and aristocratic, but not in the direct succession to power, and thus privileged but dispossessed.

*First published as 'The Rhetoric and Ethics of the Environmental Protest Movement', *Meanjin*, December 1972, pp. 427–38.

The melancholia can often be seen to be a projection of personal despair on to the cosmic screen, and its moral danger is that it so often leads to paralysis of the will. Melancholics are often charming and witty people, but the melancholia is a form of neurotic self-indulgence. They are not prophets: the prophets may foretell terrible things, but this is a conditional doom, because the emphasis with the true prophet is not on future events, but present behaviour, and 'Lest ye repent' is the heart of his cry. Paul and Anne Ehrlich have the true prophetic ring, in that the doom they predict is seen to follow only from the neglect of certain specified courses of action. Whether or not they are always right, they always give an argument, backed by a wealth of detailed information, and their passionate polemic is not cheap rhetoric (Ehrlich and Ehrlich, 1970).

The melancholic is often a manic-depressive, and in his manic phase is Utopian, with a great belief in, to take a twentieth-century example, town planning as a cure for all ills. Sometimes the despair comes of asking too much, in the manic phase, of human behaviour and human institutions. The rhetorical style of the Jacquards is 'Galgenhumor'. Here are a few examples, collected over a period of a year in the USA:

> I am an urban planner – that is to say, it is my job to rearrange the chairs on the decks of the *Titanic*.
>
> Those who think the world is coming to an end are guilty of wishful thinking.
>
> My fear is not that we will fail to adapt to the changes in our environment, but that we will succeed.

190

A cosier, small-town version is that 'The trouble with the present is that the future's not what it used to be', an epigram that charts the decline of optimism in America. That optimism was sometimes shallow, but there is no advance in replacing it with a shallow pessimism.

Gallows humour shows high courage when the doom is certain, and it offers a release we all may need at times; but it is not good for the conduct of our daily affairs, in which we need, in Aneurin Bevan's words, 'to achieve passion in the pursuit of qualified objectives', for qualified objectives are the only ones we are likely to realise, whereas the whole point of gallows humour is that we are absolved by the circumstances from further action. When a whole society becomes addicted to this mode it may lose its will to survive, and our only defence against it as individuals is patience and the courage of perseverance. Our individual contribution to global problems must always seem trivial compared with the size of the problems, but we must not be discouraged from making it. We have some choice in what we buy, and we can buy the articles with a lower environmental cost, we can live more frugally, we can exert ourselves in political protest against specific abuses. These efforts are all so petty that we can easily lose heart, especially if we are periodically frightened out of our wits by the Jacquards, who tell us that these efforts are irrelevant 'in the face of the global problems'. But it is this all-or-nothing logic that is at fault.

Categorisation such as that attempted here is always itself an oversimplification and one can easily work the analogy between the Jacobean and neo-Elizabethan melancholics too hard. It is perhaps worth insisting on dispossession as a common feature. In the case of the Jacobeans this was tied to the rise of the class of courtiers, who were generally educated, privileged, and able, but with little power and thus inadequate outlets for their abilities. In the present case dispossesion is also related to a restriction of upper-class privilege, a 'restriction' that follows from an extension of privilege in an affluent society.

The outrage of dispossession comes out very clearly in the claim, most common in crowded Europe, that a certain place is 'spoiled'. The implication is not that *no one* should be allowed to go to this place, but that no one other than one's self and a small group of like-minded people should go there. It is always others who spoil a place. This claim is often a just one, because the small group that went there before 'the invasion' usually came from a privileged class with an educated respect for landscape, and their impact was small because the group was small. Until very recently, mobility was a high and rare luxury. Today my family and I can decide at ten o'clock on a Sunday morning to leave our home near the centre of a city of three million people, for a magnificent national park some fifty kilometres away, and be there within an hour. The price we pay for this privilege is that others have it too. Places that are to remain 'unspoiled' must continue to be hard to get to, and the Forest of Arden out of reach by car. (A place may be 'spoiled' by bad design as much as by crowds. But the 'bad design' usually turns out to consist in service stations, motels and other such plebeian structures indicating mass use.)

The Revivalists

Far more common than the Jacquards, the Revivalists too have a long ancestry. There have always been those who have made use of catastrophes to induce guilt, and to show that their enemies are allied with the forces of darkness, themselves with the children of light. Earthquakes or plagues were held by some to be a Divine visitation upon our sins. Now that we know the natural causes of earthquakes and plagues, the environmental crisis (like the AIDS crisis) is an opportunity to revive a dichotomy, of the saved and the damned. To suggest that we have a range of specific practical problems before us, which we must work at with skill, patience and resolution, is as little relevant in Revivalist eyes today as it would have been three hundred years ago to suggest to their precursors that we should investigate the natural causes of plague rather than cover ourselves with ashes. But the unacknowledged goal of Revivalist rhetoric is the self-gratification of the preacher. It is relatively easy, if you have the knack, to induce guilt, fear and terror in your audience, and a gratifying exercise of power.

It is easy to find examples of Revivalist rhetoric among the environmental protesters. Ian McHarg is (or was, at the time of writing) a deservedly eminent professor of landscape architecture at the University of Pennsylvania and the author of *Design with Nature*, a book that had biblical standing for a time. Notwithstanding, he showed himself the Billy Graham of the environmental scene in an address to the Royal Australian Institute of Architects in Sydney in 1971. This talk, from a distinguished (and rather expensive) overseas guest was received with acclaim. It has been printed in several prestigious journals. It was a very bad talk, and would not be worth analysis if it had not met with such an enthusiastic reception.

Professor McHarg's paper threatened hell-fire, offered salvation, and presented an aggressively Puritanical world view. It was anti-science, anti-rationalist, anti-education. To support these charges it is necessary to quote at length. The paper, bearing the lurid title 'Is Man a Planetary Disease?', began as follows:

> I am about to discuss two values, and because I am not objective, I will excoriate the one I do not like and laud the one I like. The two values are first, that one which we have absorbed with our mother's milk, in kindergarten, high school, through theology, and so on: the *western view* of the relation of man to nature, also called *anthropocentrism*. The other is *salvation*: the ecological view of the world.
>
> The first thing I have to say about the view which is absolutely implicit in the *whole of western* culture is that it has no correspondence to *reality* and is the best guarantee of extinction. (McHarg, 1971, p. 39, my italics)

This piece of rhetoric is so empty that there is no arguing with it, but we note the two-valued logic, the promise of salvation, and the absurd oversimplifications in identifying the enemy, especially when we remember that the speaker flew by jet from America to give the talk in a centrally heated, electrically lit auditorium with sophisticated acoustic devices.

There is no single 'western view' of the relation of man to Nature. We think of the 'Great Chain of Being' which lay behind much Mediaeval thinking, and later of the Argument from Design; we think of Saint Francis of Assisi and Rousseau and Gilbert White and William Wordsworth and Charles Darwin as well as the ignorant whalers who clubbed the Galapagos turtles to the point of extinction. We remember that the sciences of biology and ecology are Western achievements, and that although Europeans hunted the whale, it is largely through their intervention that Africa still has a fair sample of its magnificent fauna. As for anthropocentrism, there is no intelligible alternative. We can regard the world only from our own point of view. We may take a long view of our own interest, or a short one, but we cannot take an insect's view, nor that of a plant pathogen, and it would be plain silly to try.

The enemy then stand up to be counted. They turn out to be the Generals, the Mad Scientists, and the Captains of Industry:

> The men who constitute the very quintessence of the planetary disease are the heads of the Defence Department, the Generals Overkill . . . They are the absolute incarnation of the planetary disease: they are pus, they threaten the survival of all mankind, and they must be chained.
>
> To the Generals Overkill must be added the manic Dr Strangeloves, men who begin life pulling the wings from flies and gravitate to bombs and high explosives and atomic weapons. In the USA they can make explosions legally underground and illegally above ground, attacking the genetic inheritance of the world. (p. 39)

How convenient that they should bear the guilt, not ourselves. These figures are bogey men, got up to frighten the children in the dark. Not all biologists begin their careers by pulling the wings off flies, although I take it (perhaps wrongly, but who knows when the rhetoric is so wild) that this sadistic image is intended to indicate a violence that is inherent in the scientists' analytical approach to nature, whereby they take things apart. It is easy to arouse feelings that are hostile to science and technology today, but this Luddite revival is dangerous and misconceived. For example, it requires sophisticated technology to measure mercury or DDT levels and indeed most environmental monitoring is highly complex. It would solve nothing to smash the machines. This is not to deny that we urgently need changes in present technological practice, including widespread recycling of resources. Such changes are beginning to be made now. They are the outcome of arduous research programs, needing the support which the anti-science movement denies them.

Next we find McHarg adopting a Jacquard world view (my rhetorical modes are not mutually exclusive):

> The world body is covered with lesions . . . The very heartland of the planetary disease is the USA. We all share the view which is implicit in all western cultures and always has been. It is very difficult to find the origins of this attitude of man to nature, but it is verified in *Genesis*. (Being a Presbyterian, I can never escape from theological questions.) (p. 40)

A good Presbyterian takes theological questions a little more seriously than this, but the oversimplification goes on at a great rate:

> The basis of our attitude to nature, whether we are Jews, Christians, agnostics, or atheists, is quite explicit in three lines. The first one is that man is made in the image of God. No atoms: no micro-organisms; no animals, save one. The implications are clear: the world consists of a dialogue between man and God, and atoms are unable to speak to God because of his interminable dialogue with man. If you covet your neighbour's wife, the Church, the priest, and society will rap you across the knuckles, but if you want to kill every single whale or fell forests of redwoods, or destroy or rape or poison any part of the natural world, you may do so, if not with the Church's blessing, at least with its consent. *There is only one moral arena: the relation of man to man.* If you so examine our cultural history you will see that this is so. There are no defined duties or rights for beasts and things. Only man has rights . . . The intent of the text is that man is given exclusive divinity, man is given dominion, and man is enjoined to multiply and subdue the earth. This one text explains all the despoliation that western man has accomplished in Asia, in Europe, and in North America. It is *the* only text you *have to know*. (p. 40, my italics)

193

The 'only text you have to know', which offers salvation or damnation in three gospel lines, is the authentic note of the Revivalist; but I prefer the Book of Genesis, a myth of such richness and complexity that it has been subject to a multitude of interpretations, although it is generally agreed that Man and Beast lived in harmony in the Garden of Eden. Much, of course, follows from the Fall, when Man was enjoined to earn his bread by the sweat of his brow. This has been inescapable, but it led to the good husbandry of the old Spanish missions in California as well as to the felling of redwoods. Doubtless the Church has much to answer for, and the sharp distinction between Man, unique in having a soul, and the other animals, all soulless, is one of the least attractive aspects of Christianity. But it is too simple to lay all the blame at the door of one text. The Romans had most of the qualities McHarg objects to before they became Christian. The role of the Church in the Western attitude to Nature has been complex: a number of advances in agriculture and husbandry were made within the walls of monastery gardens, among them Gregor Mendel's genetic discoveries, which have fed so many hungry mouths.

In the next few paragraphs McHarg's global view continues, and a 'planetary doctor' is brought on stage to announce that 'the brain is not the apex of biological evolution but a spinal tumour'. The anti-rationalism then comes thick and fast: 'I want to go into a tirade against education, which is a device for producing functional cripples. It manages to avoid teaching all the most important things' (p. 41).

One might wonder in passing what non-functional cripples would be like, but this kind of query is relevant only when words are used with care for their meaning. We then proceed to fantasy. Generals Overkill, the mad scientists and all the other baddies are shot off into space in a non-returnable rocket, and lo! we are saved: 'After they had left the Earth, if you listened very, very closely, after the reverberations, you would begin to hear the sound of healthy tissue growing, the Aeolian hum of regeneration' (p. 41). This is the organic metaphor run mad. Nature is not benign, and that Aeolian hum might well turn out to be a new and massive plague of locusts hatching in Ethiopia – to be fought with sticks, since the mad scientists presumably took their aerial sprays with them into space.

After this nothing need surprise us, not even to learn that 'Algae know about creativity, but man does not' (p. 42). My own preference is for Mozart and Einstein, but perhaps I have not met the right algae. Finally, we have to learn a prayer, and this is where we find the eagerly awaited extension of the moral arena, hitherto unreasonably restricted to the man–man relation: 'It is simply a plain prayer addressed to the world at large, revealing an understanding of the way the world works. It is addressed to the elements, to hydrogen, helium and so on. "Matter, exist for the benefit of the universe"' (p. 42). However desperate the environmental crisis, I will not say prayers to helium, and I find 'the benefit of the universe' unintelligible. Ethics is to be replaced by pantheism. As this suggestion has been argued more carefully by others, I shall leave it until later.

The Bucolics

My third, last, and least well-defined rhetorical mode is the Bucolic, annihilating all that's made to a green thought in a green intellect. It reached its nadir in Charles Reich's *The Greening of America*, which offers the following:

> The extraordinary thing about this new consciousness is that it has emerged out of the wasteland of the Corporate State, like flowers pushing up through the concrete pavement. Whatever it touches it beautifies and renews: a freeway entrance is festooned with happy hitchhikers, the sidewalk is decorated with street people, the humorless steps of an official building are given warmth by a group of musicians. Every barrier falls before it.
>
> The new consciousness is sweeping the high schools, it is seen in the smiles on the streets. It has begun to transform and humanize the landscape. When in the fall of 1969, the courtyard of Yale Law School, that Gothic citadel of the elite, became for a few weeks the site of a commune, with tents, sleeping bags and outdoor cooking, who could any longer doubt the clearing wind was coming? (Reich, 1970, p. 3)

The image of the flowers pushing up through the pavement is sinister in a way hardly intended by Reich. It happened in London after the Blitz and it marked the destruction of a city with all the suffering that entailed. As for the sleeping bags outside the Yale Law School, it would be well if all our troubles could be solved by camping (even high camping). This rhetoric has in common with that of the Revivalists the offer of simple solutions for complex problems. The language and logic are so childish that Henry Fairlie (1971) wrote a brilliant review of *The Greening of America* under the title 'The Practice of Puffers'. He coined a word for such language: 'Babyspeak', by analogy with George Orwell's 'Newspeak'.

The pastoral idyll is a recurrent literary mode, usually a by-product of urban malfunction. Many writers have recoiled from the political and social instability of

194

the city to the good husbandry of the countryside, or the purity of the wilderness. But Reich wants more than this: the city must *be* the country. This odd theme is presented visually in many public lectures on the 'environmental crisis'. In the USA a common rhetorical device is to project side by side a breathtaking colour slide of, say, the High Sierras, and a freeway interchange or gas-station and parking lot in Los Angeles. These are presented, often without comment, as if they were alternatives, which of course they are not. One cannot buy gas in the High Sierras, and it is foolish to go rope-climbing or backpacking in downtown Los Angeles. It may come as a shock to some to learn that the Sierras and Los Angeles are in high degree complementary. The prime reason that so much of the Sierra Nevada is still near-'virgin' country is that the city jealously guards its water-catchment. Without Los Angeles, the Sierras would probably by now be mined and logged and ranched to death.

The agrarian myth has been potent in the USA, as in Australia. Perhaps American society might have been happier had it remained an agrarian society; but whatever might have been, it is now urban and industrial. Many of the major environmental problems, both in America and Australia, are in the cities; but in neither country is there a full emotional grasp of this reality, and the conservation issues that inflame the passions still tend to be the despoliation of the countryside, the threatened extinction of the red kangaroo or the bald eagle, dam-building at Hetch Hetchy or Lake Pedder. These are proper concerns, fought in major part by city people, but in the meantime they turn their back on the city itself, which is one reason why American and Australian cities lack the civility of their European counterparts. The contrast in attitudes comes out nicely in cigarette advertising: Australian and American cigarettes are smoked by lean horsemen, eyes slitted against the glare and dust – but Benson and Hedges cigarettes are a love-offering to a girl in a city restaurant.

195

Assigning the blame

The rhetoric of the Marlboro countrymen or Madison Avenue Bucolics is usually tinged with animism, which is also the creed of some of the Revivalists. It was expounded at length in 1967 by Lyn White, a Professor of History at the University of California at Los Angeles, in a paper entitled 'The Historical Roots of Our Ecologic Crisis'. This much-admired paper was on the reading list in the 1970s for all those proliferating courses on the environment in the USA. It set the fashion, followed by McHarg and many scientists who are short on history, of blaming all our problems on 'Judeo-Christian teleology', because 'God planned all of this explicitly for man's benefit and rule: no thing in the physical creation had any purpose save to serve man's purposes' (White, 1967, p. 1205). This is not the view of the psalmists: even the first chapter of Genesis, to which Professor White's knowledge of the Judaeo-Christian tradition seems to be restricted, gives the lie to so simple a view: 'God created great whales . . . and every winged fowl after his kind: and God saw that it was good' (Genesis 1:21) . It was good before man, they were good after their own kind, and although all this was for man's use, this was never their only 'purpose'.

Next we learn that 'by destroying pagan animism, Christianity made it possible to exploit nature in a mood of indifference to the feelings of natural objects' (p. 1205), and that 'To a Christian a tree can be no more than a physical fact. The whole concept of the sacred grove is alien to Christianity and to the ethos of the West . . . "When you've seen one redwood tree, you've seen them all" . . . For nearly two millennia Christian missionaries have been chopping down sacred groves which are idolatrous because they assume spirit in nature'. This is so confused it is hard to know where to begin. The 'concept of the sacred grove' as White conceives it owes at least as much to the Greeks and their dryads, a part of Western cultural history, and to Wordsworth and Naturphilosophie of the nineteenth century as it does to New Guinea tribesmen who think that trees must be placated. As for trees as 'no more than a physical fact' – I suppose if we want to make such statements, they are a biological rather than physical fact, but what then? What more can they be? And it certainly does not follow from such a view that 'when you've seen one redwood tree, you've seen them all'. I can, I suppose, concede that a tree is no more than a biological fact – although I find it hard to think of circumstances under which I should want to make such a remark – but this does not stop me from liking trees, nor from admiring a redwood grove more than a single freestanding redwood, nor from trying to stop people from cutting trees down unnecessarily, and I doubt that any of these attitudes could be reinforced if I thought there were spooks in trees. Incidentally, nearly all of the major reforestation programs in the Middle East, around the Mediterranean, in Africa and elsewhere have been the work of Europeans, and the Jews have done wonders in Israel.

Lyn White doesn't really think there are spooks in trees either, but he wants us to pretend there are. 'More science and more technology are not going to get us out of the present ecologic crisis until we find a new religion, or rethink our old one' (p. 1206). Saint Francis of Assisi is to be our guide, and we should recognise Brother Ant and Sister Fire, praising the Creator in their own ways as Brother Man does in his. This is hardly novel; it is the view of the psalmists. But what then? A fable. 'The land around Gubbio in the Apennines was being ravaged by a fierce wolf. Saint Francis, says the legend, talked to the wolf and persuaded him of the error of his ways. The wolf repented, died in the odor of sanctity, and was buried in consecrated ground' (p. 1207). Well, bully for Saint Francis; but most of us are not much good at talking to wolves or exhorting the birds, as White later acknowledges. Thus this fable can have no real application, and although White proposes Francis as a patron saint for ecologists, the new religion will not help us. Brother Ant, perhaps. Brother Streptococcus?

It is now in fashion to think ill of our own species. I read in a recent book that 'If I were God I might well prefer swallows, squirrels and butterflies to man' (Allsopp, 1972). I suppose this to mean that, in some moods, the author prefers them, and the rest is anthropomorphic extension of a perverse variety. If this is a new religion, it would be a very odd one for a member of our species. But I do not think we need a 'new religion', least of all if it turns out to be a romantic, nineteenth-century animism. There is, surely, a case for the extension of our ethical conscious-ness. McHarg complained of the view that 'there is only one moral arena: the relation of man to man'. This plaint goes back to Aldo Leopold (1949), who argued the case carefully years ago. Leopold discusses the extension of ethical criteria through history to new fields of conduct 'with corresponding shrinkages in those judged by expedience only'. Odysseus, for example, applied ethical standards to his relations with his wife but not his slaves. They were property. Land is still property:

The land-relation is still strictly economic, entailing privileges but not obligations.

The extension of ethics to this third element in the human environment is, if I read the evidence correctly, an evolutionary possibility and an ecological necessity. It is the third step in a sequence. The first two have already been taken. Individual thinkers since the days of Ezekiel and Isaiah have asserted that the despoliation of land is not only inexpedient, but wrong. Society, however, has not yet affirmed their belief. I regard the present conservation movement as the embryo of such an affirmation. (pp. 202–3)

The custodial view

I would agree that the despoliation of land is wrong, and Leopold goes no further than this. But why is it wrong? Surely because it is an infringement of the rights of later generations, and not because the land has some mystical rights of its own. We must take the custodial view. From time to time we are told that we should learn to respect wild things 'for themselves', and not merely as potential items of use. With this, too, I can agree. Anyone who regards a redwood tree as no more than so much merchantable timber is an impoverished soul. But I can respect redwoods 'for themselves' and readily concede that they have a life of their own that is different from mine without supposing that they have feelings. It is *my* feelings that matter, and if they are mean, it is I or my grandchildren who suffer.

Our relations with animals are more complex. A community that supports a Society for the Prevention of Cruelty to Animals and upholds it in the law courts clearly supposes that animals feel. Moral obligations to animals are generally recognised, I think, and more so in Western than in most non-Western cultures. For example, it seems to me a moral commonplace of our society that it is wrong to keep a dog and not exercise it properly. There is also a feeling, deeper than this, that we degrade animals by locking them up in zoos. There is a general recognition of the beauty and power of animal lives other than our own, and that is best displayed when the animals live their own lives untouched by man (*Born Free* was a best-seller, Professor White). Many individuals are cruel to animals, or needlessly destructive, but the mere use of the word 'cruel' in this context shows that society does take an ethical view of the relation, although it does too little to enforce that view. But no matter how much we respect the feelings of animals, we are still going to eat them and/or the plants, and this paradox is much older than Judaeo-Christianity.

The superb Palaeolithic paintings of bison and other animals at Lascaux and Altamira show a great knowledge of and sympathy for the animals depicted, but this sympathy is not sentimental pantheism. Palaeolithic man imitated his prey in painting and in ritual dances. All primitive hunters know a tremendous amount about the animals they hunt, and they respect their cunning and skill – indeed to be a successful hunter you have to learn to think like your prey. The fruit-eating primates, from which the hunting apes diverged in the course of evolution, did not kill other species for food, but it is unlikely that they held them in reverence either. Presumably they knew very little about them. Our reverence for life derives, in this sense, from being the killer apes. But this is a far cry from romantic pantheism. If I seem to labour the point it is because the pantheism urged by rich North Americans and Australians is largely make-believe, and often self-indulgent. The wilderness cult is full of it: the return to simplicity – by canoeing down the Colorado or the Franklin, for example – is for a week, and based on a logistic that is anything but

197

simple. And it all diverts attention from 'the relation of man to man', the moral arena in which terrible things are done daily. Black militants in the USA were quick to point out that the 'ecology jag' has diverted attention from racial injustice.

The custodial view of nature has long been inherent in good agricultural practices. The agrarian resources of Europe have been conserved for centuries. The custodial view is weakest in newly settled lands, and needs all the encouragement we can give it through legal and tax support, but it is not new. What is new is the extension of the custodial view to our industrial activities, to make sure that renewable resources are renewed and that non-renewable resources are kept in cycle. This view is now accepted in principle by all thinking people, and the formidable problem for the future is to put it into practice. The State of Oregon passed a law (in 1972) that all drinks be sold in returnable containers, and Minneapolis began to recycle its garbage. Most Australian cities now do so. It is dull stuff compared with religious conversions, but it points the way ahead.

There is perhaps one other novel element in the expansion of the custodial view: we are becoming a great deal more conscious of the unintended consequences of intentional actions. Any act has physical consequences only part of which stem from the intention. These may be trivial: they may not. The men who invented DDT did not intend to weaken the egg-shell of the American eagle any more than Professor McHarg intended to contribute to atmospheric pollution by flying from Philadelphia to Sydney. But both things happened. Our managerial responsibilities now extend to care for the atmosphere, to oceans, the biosphere, and thus our custodial role is on a new scale. This is directly contrary to the view of White and McHarg and Reich that all our troubles arise from our pretensions to dominion over Nature. We were expelled from Eden, and the gates are guarded with a flaming sword. Or to put the point plainly and without rhetorical flourish, it is rare to find simple solutions to complex problems, and we are not likely to do so in the present case. Rhetoric that induces despair and hence inaction will not help us. Contempt for and alienation from Western society may be productive to the extent that it produces critical self-awareness but many go beyond that. Western society has, historically, been self-reforming in a degree unmatched by any other, and it is not dead yet. It is counter-productive to destroy the faith of the young in their own society. Extreme rhetoric is socially divisive, and it is sure to provoke ecologic backlash. It debases the quality of public debate. For all these reasons, it should be eschewed.

Postscript

Although this essay was written in 1972, extreme rhetoric is hydra-headed. Lop one off: another grows in its place. As for global conservation; some things are better, others worse. Although the recycling of garbage is now fairly common in Australia, prompted largely by the cost of landfill and cartage, Europe has long given up husbanding its agricultural resources. *La plus ça change . . .*

References

Allsopp, Bruce, 1972, *Ecological Morality*, London: Frederick Muller.

Ehrlich, Paul R., and Ehrlich, Anne H., 1970, *Population, Resources and Environment*, San Francisco: W. H. Freeman.

Fairlie, Henry, 1971, 'The Practice of Puffers', *Encounter*, vol. 37, no. 2, pp. 3–13.

Leopold, Aldo, 1949, *A Sand County Almanac*, London: Oxford University Press.

McHarg, I., 1971, 'Is Man a Planetary Disease?', *The Architect*, Western Australian Chapter of the Royal Australian Institute of Architects, September, pp. 39–44, reprinted from *Journal of the Royal British Institute of Architects*, July 1970.

Reich, Charles, 1970, *The Greening of America*, New York: Random House.

White, Lyn, 1967, 'The Historical Roots of Our Ecological Crisis', *Science*, March, vol. 115, series 37, no. 67, pp. 1203–7.

21 The perfectibility of Nature: a review of John Passmore's **Man's Responsibility for Nature***

Ecological problems and problems in ecology

The publication of John Passmore's book is a major intellectual event. It is in three parts. The first consists of two historical chapters, looking at the complex set of Western attitudes towards Nature, and their sources in our past. The second consists of four chapters, one on each of the major ecological problems of our day: pollution; the depletion of natural resources; the destruction of species; overpopulation. The third part, entitled *The Traditions Reconsidered*, consists of a single chapter, 'Removing the Rubbish'; it begins with John Locke, who sought 'to be employed as an under-labourer in clearing ground a little and removing some of the rubbish that lies in the way to knowledge' (Locke in Passmore, 1974, p. 175), 'and by way of knowledge, to effective action', Passmore adds as Locke did not. When I first read Locke's phrase some years ago, I was delighted and mildly outraged by the presumption of such modesty. The master-builders of the commonwealth of learning acknowledged by Locke were 'the great *Huygenius* and the incomparable Mr. *Newton*, with some others of that strain' (Locke, 1706, p. xxxv); it was in relation to them that Locke saw himself as an under-labourer. But none of the scientists of his day was likely to have written the 'Essay concerning Human Understanding', and Locke knew it. A part of his prefatory intention was to indicate the kind of task on which an analytical philosopher might usefully be employed, and he was thus an early spokesman for the anti-metaphysical, analytical bias of the British philosophical tradition. Hume expanded Locke's metaphor by declaring a bonfire:

> When we run our libraries, persuaded of these principles, what havoc must we make? If we take in our hand any volume; of divinity or school metaphysics, for instance; let us ask, *Does it contain any abstract reasoning concerning quantity or number?* No. *Does it contain any experimental reasoning concerning matter of fact and existence?* No. Commit it then to the flames: for it can contain nothing but sophistry and illusion. (Hume, 1777)

*First published as 'The Perfectibility of Nature', *Meanjin*, vol. 33, no. 3, September 1974, pp. 249–55.

Two intricate natural geometries, perfected for their function of entrapment? The birdsnest fern (*Asplenium nidus*) captures rainfall and channels it towards the roots; the spider web captures nutriment. COLIN TOTTERDELL

Passmore is too much the historian to wish to burn books, even the bad ones, but he is firmly aligned with those British philosophers who eschewed the building of systems, and in our own day came to see their task as a therapeutic 'analysis of discourse'. Thus, at least, the program: but as the tools of analysis have become sharper, the range of discourse has shrunk, and has tended to become one of professional set-pieces.

John Passmore proposes a new range of application for the analyst's skills: the discourse of the environmental debate. Like Locke, he hopes to be useful: 'It will possibly be censored as a great piece of vanity or insolence in me to pretend to instruct this our knowing age' (Locke, 1706, p. xxxiv) says the one. And the other: 'If it contributes slightly to clearer thinking about the problems which confront us, I shall be more than satisfied' (Passmore, 1974, p. vii). Locke was not ashamed of aiming to be useful: 'methinks it savours much more of vanity or insolence to publish a book for any other end'. Passmore also shows Locke's mock-innocent surprise at the philosophical preoccupations of his contemporaries: for example, of John Rawls' *The Theory of Justice*: 'He does not so much as mention the saving of natural resources. (How rare it is for moral philosophers to pay any attention to the world around them!)' (p. 86). Finally, Passmore allies himself with Locke in having a high regard for science as a way of knowing; 'ignorance is one of the most potent obstacles to our solving our ecological problems, an ignorance which only science can dispel' (p. 176); although he also recognises that science is in need of new directions, and that science is not philosophy. He draws a distinction early in the book (p. 43) between 'ecological problems' and 'problems in ecology'.

> A problem in ecology is a purely scientific problem, arising out of the fact that scientists do not understand some particular ecological phenomenon, how, for example, DDT finds its way into the fat of Antarctic birds. Its solution brings them understanding. An ecological problem, in contrast, is a special type of social problem.

The terminology here is a little hard to remember, which is likely to prevent its coming into general use, but it is a major distinction, central to the thesis of the book, and we shall return to it.

The framework of the thesis is constructed in the first two chapters, and the last. It consists in historical analysis of Western traditions and attitudes to Nature, to prepare us to assess the cry of the environmental extremist for 'a new morality, a new religion, which would lead him to believe that it is *intrinsically* wrong to destroy a species, cut down a tree, clear a wilderness'. But a morality, a religion, is not the sort of thing one can simply conjure up, and Passmore quotes Hotspur to ridicule the new Welsh wizards:

> Glendower: I can call spirits from the vasty deep
> Hotspur: Why so can I, or so can any man,
>   But will they come when you do call for them? (Passmore, p. 111)

Western culture is far from monolithic, and, by patient quarrying, Passmore can find much in our inheritance that is relevant to our current concerns. His regard for the social fabric is one of the most valuable features of this valuable book, evinced again and again in his regard for continuities, and in assessing the social cost of proposed reforms:

there are limits to what we as a generation can do, and limits, too, to what we ought to attempt; our concern is not with the numbers of posterity as such but with their sensibility, their civilization, their happiness. To surrender our freedom, to abandon all respect for persons, in the name of control over population growth is to make sacrifices which our proper concern for posterity cannot justify. (p. 170)

The history of some key concepts

It has been fashionable in the last few years, especially in the United States of America, to attribute our current ecological problems to fundamental deficiencies in the Western tradition, which are part of its very fabric because they are central to Judaeo-Christian teleology, the view that God so planned the world that 'no item in the physical creation had any purpose save to serve man's purposes' (White, 1967, p. 1205), as discussed in the previous chapter. We noted that Ian McHarg, who has done so much for standards of environmental design, has been guilty of giving popular currency to this view in extreme form: anthropocentrism is 'absolutely implicit in the whole of western culture', he says, and he has therefore urged that we should reject that culture *in toto*, a mood that is dangerously attractive to the young. As historical analysis, this is deficient: it has some truth, some falsity. Passmore sets himself to sift the two, and to reject the conclusion.

First, Passmore rejects the conflation inherent in the phrase 'Judaeo-Christian teleology' by showing that the Old Testament does not set up an unbridgeable gap between man and his fellow creatures, and that its uncompromising theocentrism leaves no room for anthropocentrism: Nature was seen to exist, not for man's sake, but for the greater glory of God. The anthropocentrism of our culture comes rather from the Greeks – it is in Aristotle, for example: 'plants are created for the sake of animals, and the animals for the sake of men'. Passmore therefore concludes that the critics of Christianity are right in that 'Christianity has encouraged man to think of himself as nature's absolute master, for whom everything that exists was designed. They are wrong only in supposing that this is also the Hebrew teaching; it originates with the Greeks' (Passmore, p. 13). It was transmitted by the Stoics, and contested by the Epicureans, who saw men as not only using but being used by the living things that surround them, a view that was to regain currency after Darwin.

Passmore is a good companion and guide through the complexities of intellectual history. To summarise crudely what he displays subtly, history has been neither unitary nor static, and the Stoic-Christian doctrine that 'everything is made for man's sake' was not itself enough to provoke, or be used to justify, a scientific-technologic revolution. That had to await the infusion of a 'Pelagian humanistic attitude to man, which sees him not as essentially corrupt but having the duty to create, by his own efforts, a second nature' (p. 20). This 'duty to create' has found diverse expression, ranging from the dream of 'perfecting' nature, which led Capability Brown to actualise the capabilities of a site, to the ruthless transformations of nature by technologists to serve man's supposed ends. Even science itself has not been monolithic in its view of Nature. Passmore contrasts the Baconian-Cartesian approach with that of the naturalists Ray and Linnaeus, both of whom were fascinated by the intricacy of Nature, Linnaeus by 'the interplay characteristic of an ecological system for which the language of means and ends is totally inappropriate'. Descartes, by contrast, saw all animals other than man as incapable of

203

sensation (even of pain), 'mere machines or puppets'. His emphasis as a geometer and physicist was 'not on the diversity of forms but on the uniformity of laws. The qualities which make nature so attractive, notably colour, it denied nature to possess' (Passmore, p. 22). Thus Passmore makes us aware of, although he does not explore, some of the diverse elements that go to make up that complex known as 'science', which is as diverse as the culture that gave rise to it. The strands in science that run from Lucretius, and also the 'naturales' of Aquinas through Linnaeus to Darwin and the ecologists and field scientists in our own day, are fit subjects for an intellectual historian of Passmore's calibre.

The function of the two historical chapters is to suggest that the West has viable elements within its diverse traditions, and that these should be identified and built upon. By sweeping the board for a new beginning, there is no guarantee that we shall get a better society than we now have, and good reason for fearing that we may get a worse. The second chapter is therefore devoted to constructive elements in our tradition, and two of them are discussed in some detail: the notion of 'perfecting' nature, touched on above; and that of the good steward, a tradition that demands from man an active concern for the well-being of the estate of which he is trustee, a tradition more likely, in Passmore's view and mine, to lead to care for the environment than the quietism of the Buddhists, in which salvation is to be achieved only by freeing oneself from every kind of earthly bondage. The example is not quite fair, but is nevertheless worth giving, that nowhere 'is ecological destruction more apparent than in today's Japan, for all its tradition of nature-worship' (p. 176).

Pollution, conservation, preservation

It would do great violence to the argument to attempt here to give the substance of the four analytical chapters, and I shall content myself with a few samples, and direct the reader to the original. In effect, Passmore is a decontamination squad, defusing loaded terms. Of natural processes, for example, so often, so vacuously, assumed as benign, on the grounds that 'Nature knows best': 'the "natural" is not necessarily harmless, let alone beneficial to man' (p. 47). He reaffirms what should surely be obvious, that in so far as ecological problems can be solved only with the help of scientific study and technological invention, they can be solved only within the Western rational tradition. 'Mystical contemplation will not reveal to the chemist the origins of the Los Angeles smog' (p. 49). He offers for consideration the thought that particular industries, intent on maximising profits not only will but *should* emphasise the costs, as distinct from the benefits, of anti-pollution measures, 'if a society is to arrive at the optimal situation in which pollution is reduced to a tolerable level at the minimum cost' (p. 61). He assesses the relative merits of different kinds of anti-pollution measures, and takes into account, as few do, their social cost: 'If our legal system is not entirely to break down, there is good reason for believing that we need to reduce rather than to increase, as we have steadily been doing, the number of actions that are accounted crimes' (p. 70).

The chapters 'Conservation' and 'Preservation' are correctly titled, but not in accord with current usage. Passmore uses 'conservation' to cover the saving of natural resources for later consumption: the activities of bodies such as the Australian Conservation Foundation in saving species and landscape from destruction are those of preservation, since they exclude the sense of consumption in the future. The chapter on conservation, the husbanding of resources, is in some ways the nearest to traditional philosophy: it might be helpful to borrow the distinction used by scientists, and to speak of 'pure' and 'applied' philosophy. Most of this book is then applied philosophy, but this chapter is 'pure', in that it examines a philosophical puzzle. Although it is not always clear in any given case that future generations will *want* what we choose to save for them – their needs may be different – they clearly will have needs, and Passmore accepts a responsibility to posterity (not by writing a blank cheque, however: he would accept no duties towards a totalitarian world state, nor would I). The interesting philosophical question is how such a responsibility might be justified, other than as Divinely imposed. Passmore reviews the alternatives.

He brings out conflicts inherent in the preservation movement, and notes, for instance, that by putting roads into its national parks, the US authorities chose 'to put recreation first, the preservation of wilderness second – and therefore nowhere' (p. 104). The cant phrases of popular ecology are subjected to useful scrutiny. It is true that 'it is never possible to do only one thing at a time', but it is not true – in any significant way – that 'everything depends on everything else', and neither precept is a justification for quietism, the prescription to quit 'interfering' with nature. We have no choice but to live dangerously. Passmore notes that the 'ecological niches', now such favourites with the preservationists, carry one back to the 'Great Chain of Being' and the plenum of the Middle Ages, thus highlighting the curiously static, non-evolutionary character of popular ecology. He points out the illicit extension of the word 'community' in some debates. We form a community with plants, animals and soils, in the sense that a particular life-cycle will involve all four. In the sense that belonging to a community generates ethical obligation, we do *not* belong to the same community. 'Bacteria and men do not recognize mutual obligations, nor do they have common interests' (p. 118).

205

Population growth

And so the useful work of sifting and clarification goes on. The last of these four chapters, 'Multiplication', on population growth, was clearly the hardest to write, and Passmore begins to flag. Because it is clear that at some point in the future population growth must be arrested, he is obliged in this chapter to argue for a particular proposition, to examine the many impediments towards adopting such a policy, and the alternative means of implementing it. This is too much for one chapter, yet it is necessary to the structure of the book. I find as little to disagree with in this as I do in the book as a whole, but it is likely to provoke most resistance – which is part of the problem.

Passmore's humanist values and quiddities

One of the pleasures of this book is that it bears witness to the values which the author holds. Without obtrusive display of personality, it is a personal statement, not an impersonal voice in a void. Passmore speaks to the world, but from a Canberra study. It would be no great interference with nature to dispense with bush flies from a bush picnic, we are told. 'Anyone who lives on the inhospitable shores of Australia will certainly appreciate' (p. 186) John Stuart Mill's warnings against the sentimental nature-lovers. It is easier, he laments, for a professor in London to renounce jet air travel than it is for a professor in Canberra. So this voice is unselfconsciously *placed*. It is also unrepentantly middle-aged and academic. 'Man, it must certainly be recognized, has no tenure in the biosphere', we are told (p. 184); 'annual production is not the best test of efficiency – whether what is in question is a farm or a professor' (p. 121). If Western civilisation is as bad as its critics claim then 'I should have no policy to advocate except mass suicide. The alternative – the hunting savage – arouses in my heart no enthusiasm whatsoever' (p. 181). Nor is he much moved by the currently fashionable hostility to the city as destroying all individuality: 'Why, one wonders, are the streets of London and Boston, Tokyo and Sydney, awash with refugees from the narrow-mindedness, the conventionality, the constant surveillance of small towns?' (p. 53).

The voice is urbane, cosmopolitan; the landscapes of Kyushu and Nara are admired along with those of Tuscany. And yet we are also treated to some companionable prejudices, such as that anti-American wit peculiar to the senior common rooms of British and colonial universities: 'the United States – which has taken over from Oxford as the home of lost causes' (p. 61); 'In many fundamental respects, America is a very old-fashioned country' (p. 184); 'No other sort of greatness seems to be at all linked with numbers', he adds, returning to the fray, 'the United States, to say nothing of the Soviet Union, is by no other standards so great as Athens or Elizabethan England' (p. 148), a remark not wholly consistent with a central thesis of the book that the tradition of Western science is perhaps the greatest of man's achievements 'nor the fact that most of the scientists who have ever lived are alive in the last decades of the twentieth century and living in California'.

There are other trivial inconsistencies of tone which give flavour to the book, for instance a breadth of historical perspective that begins with the casual remark that 'the Greeks feared above all else the "growth" which has latterly become our ideal' (p. 158); this permits him a calm assessment of infanticide as a means of population control, and not necessarily the worst. We then move from these heights to fight again the rationalist battles of the University of Sydney in the 1940s: of James Macauley's attack on the Reverend Malthus for 'killing conscience with arithmetic', we are told that

> if we find their argument puzzling in their insensitivity, we must recall that the Roman Catholic Church has never placed much store on the reduction, as distinct from the relief, of human suffering. Their 'conscience', so much is clear, is very different from mine. (p. 131)

Nevertheless, this is central to Passmore's humanist stance: 'in my ethical arguments, I treat human interests as paramount' (p. 187). It is a humanism enriched by responsible love: 'to love is amongst other things to care about the future of what we love' (p. 88), and by a reverence for life that steers well clear of mysticism, 'the sort of reverence one feels for a great building, a great work of art' (p. 124). These offer some comfort. 'Only if men see themselves', he writes, 'for what they are, quite alone, with no one to help them except their fellow-men, products of natural processes which are wholly indifferent to their survival, will they face their ecological problems in their full implications' (p. 184). Yet there is a sustaining past. Like Edmund Burke, Passmore has a sense of continuity, and if our institutions are vulnerable – 'For not only are man and nature fragile, so too is liberal civilisation' (p. 183) – yet they are also our support.

This is not wholly a sombre work, despite the comment that 'if in the course of writing this book, I have ever felt cheerfulness breaking in, a moment's reflection on the plight of the developing countries has been sufficient to dispel it' (p. 193). The author is humane, compassionate, reasonable, witty, wise – the kind of company in which I should like to face the future, however bleak. There is more at stake than survival.

Postscript

As I said in the Prelude, our economy, ecology, society, even our language is affected by conservation issues.

References
Hume, David, 1777, *A Treatise of Human Nature*, ed. L. A. Selby-Bigge, Oxford: Clarendon Press.
Locke, John, (1706) 1965, *An Essay Concerning Human Understanding*, Everyman edition, London: Dent and Sons.
McHarg, Ian, 1969, *Design with Nature*, Garden City, New York: Natural History Press.
McHarg, Ian, 1971, 'Is Man a Planetary Disease?', *The Architect*, Western Australian Chapter of the Royal Australian Institute of Architects, September, pp. 39–44, reprinted from *Journal of the Royal British Institute of Architects*, July 1970.
Passmore, John, 1974, *Man's Responsibility for Nature: Ecological Problems and Western Traditions*, London: Duckworth.
White, Lyn 1967, 'The Historical Roots of Our Ecological Crisis', *Science*, March, vol. 115, series 37, no. 67, pp. 1203–7.

VI

Sharing and caring: ecological frameworks

Tread lightly on this land

The four chapters in this section consider what are commonly understood as 'conservation issues' in Australia, in addressing the continent as 'biophysical fact'. The first, Chapter 22, Biological pollution, is a little faded, but necessary to the theme of this book. It was written, at the editor's invitation, for an American scientific journal, and the language therefore reflects the discourse generally favoured in such journals at the time. It is faded because the data are more than ten years old; this, alas, is not critical, because extinctions continue to take place, and new 'pests' are introduced. Attempts at biological control also continue; one such is the experiment with the control of rabbits using the calicivirus, another, the test release of bass, an indigenous  Australian freshwater fish, into the water storage at the Fitzroy Falls in New South Wales. The hope is that they will break the carp breeding cycle, the carp being so numerous that they disturb bottom sediment, in-crease turbidity and affect water quality. So the details change, but the story remains the same.

What the essay lacks is passion and immediacy. The list of extinctions at Kinchega National Park tells the story in its own way, but does it prove upon the pulses of non-biologically inclined Australians the enormity of that loss? Geoff Park, a New Zealander, sets out to do that for his islands and people in *Nga Uruora: The Groves of Life*, which I reviewed for *Meanjin* in 1996, and discuss in Chapter 25.

The second essay, Chapter 23, The lie of the land, was written specifically for this book. I had considered using the title for this book as a whole until a book with the same title appeared recently, by Paul Carter (1996). Its appeal for me is the triple sense. First, there is the literal, topographic one; second its metaphorical application, as in 'sussing out the lie of the land' on Saturday, to see whether mum or dad would be good for the wherewithal to go to the pictures. The relevance of this usage is that it applies to situations in which a direct, frontal attack is risky without the skilfull reading of a range of indirect signs and signals first. The third meaning is the oxymoron: the land does not, could not, tell lies. But it seemed to do so to the men who named Useless Loop, Mt Disappointment, Point Deception. If they were deceived, it was because of false expectations.

I am no ecologist, but this essay is necessary to the theme, to show that many aspects of the Australian 'natural' scene that we tend to see as isolated snippets of information are interdependent. It is one of the simplifications of ecospeak to assert that 'everything depends on everything else'; but there is, nevertheless, an intricate web. It is not the general assertion that is of interest here, but the specific examples and pathways. Example. Australia is exceptionally rich in parrots. Why? Many Australian plants have hard, woody seed capsules because of adaptation to seasonal aridity, low-nutrient soils and wildfire. Parrots can be seen as flying nutcrackers. Australia does not, however, have any woodpeckers. Why? Well, read the essay.

All professional ecologists are aware of a multitude of such correlations, but not many of them communicate them in ways that are both scientifically accurate and vividly arresting to a general public which I think is avid for such knowledge. Happily, there are some fine exceptions, such as Geoff Park, noted above; and Richard Braithwaite and Tim Flannery, who features in Chapter 24, Eating the future. This is a review of Flannery's book, *The Future Eaters*. When I wrote the review, I had no idea that the book would be a best seller, but I am immensely heartened by its success, partly for the author's sake (because he writes with that 'immediacy and passion' that used to be seen conventionally as inimical to sound science), but even more, because it is incontrovertible evidence that Australians want books such as his. Although *The Future Eaters* changes the terms of the debate, its focus differs from my preceding essay; his framework is historical, and the book is driven by a single passionate theme, that our species is, indeed, a future eater, unrestrained in its resource consumption. Our 'great leap forward' from being merely one species among many to the dominant species, was achieved by this characteristic, when it was released from the pressures of co-evolution by crossing Wallace's line from Bali to Lombok. The same characteristics that ensured its success will ensure its destruction unless we mend our ways.

Chapter 25, Felling the 'Groves of Life', is a discussion of resource consumption in New Zealand, and especially on the destruction of its incomparable lowland forests.

References

Carter, Paul, 1996, *The Lie of the Land*, London: Faber & Faber.

22

Biological pollution*

Australia has a very different history from that of other industrialised countries. Industrial pollution is real in and around our larger cities, but it is relatively localised, and our control measures are similar to those applied in other industrialised countries.

The two remarkable features of Australia are that it is a large-scale island ecosystem, and that the impact of Europeans, and of the plants and animals they brought with them, has taken place within two hundred years. Australia is the extreme example of an isolated ecosystem by world standards. The special vulnerability to disruption of island ecosystems is well known to biologists. Australia is the only such system on a continental scale, and for that reason, some aspects of man's relationships with the biosphere stand out with startling clarity.

The consequences of this isolation are apparent even to the most casual visitor. Until recently, everyone who arrived by airplane was subject to an insecticidal spray. Quarantine is still strict, and great precautions are taken to avoid stock diseases such as 'blue tongue' in sheep, foot and mouth disease, and swine fever. An occasional sparrow, starling or Ceylon crow arrives by boat at Fremantle, the major port of south-western Australia. The sparrow and starling are now common in eastern Australia, but they have not yet reached the west, so odd escapees are pursued with rifles until they are shot. Some of these precautions have an air of comedy, but the reality that lies behind them is grim. Australia has proved extraordinarily vulnerable to introduced diseases, pests and weeds. Air travel, by increasing the ease and frequency of arrivals from overseas, has greatly increased the dangers.

Differences from the northern hemisphere

The lands of the northern hemisphere are the great land mass of Eurasia, which is essentially continuous with Africa. North America has been linked across the Bering Strait with Eurasia, and both flora and fauna are closely related. South America was isolated during much of its evolution, but was linked by a land-bridge with North

*First published as 'Biological Pollution in Australia', *Resource Management and Optimization*, vol. 2, no. 3, 1983, pp. 293–305.

America in the late Tertiary, which enabled an invasion of placental mammals. In evolutionary terms, the settlement of Australia by Europeans in the late eighteenth century is comparable in the magnitude of its biological effects with this major geological event, the linking of the Americas in the Pliocene and Pleistocene, over a period of some five million years (beginning about 6 million years ago). That is rapid in geological terms, but similar changes are taking place in Australia over a period of a few hundred years. By the time-scale of our own lives, the speed of biological changes taking place and their magnitude are not fully apparent. For instance, the State of Victoria has lost about 10 per cent of her indigenous plant species in the last hundred years (Turner, 1966); 25 per cent of its total flora is made up of introduced species – and that means plants that 'grow wild', not those of our gardens. In biological and evolutionary terms, that is an enormous and rapid change, far more spectacular than the extinction of the dinosaurs, itself a sudden event geologically, but still one that ran over millions of years. Yet, to human beings, events that take place over three generations seem slow.

There has been a long history of mutual adjustment, which is in large degree taken for granted, between man and the biosphere in Africa, Eurasia and the Americas. All of the crop plants and all of the domesticated animals come from this vast linked global land mass. None of them comes from Australia (the *Macadamia* nut, grown commercially in Hawaii, is a rare exception). Cereal crops are a modification of natural grasslands of this semi-global land mass. The herds of sheep and cattle are a refinement of the herbivores that grazed those grasslands at the beginnings of human history. Not so in Australia. Australian wild grasses have not undergone millennia of selective breeding to yield cereal crops. Native herbivores have not been domesticated, although the kangaroo is the ecological equivalent of the deer and cattle of Eurasia. The weeds, the diseases, the pests of Eurasian agricultural systems are also essentially derivatives of natural ecosystems; although man has greatly modified those ecosystems, they are not wholly alien introductions.

But the behaviour of introduced plants, animals and pathogens in Australia is essentially unpredictable. The natural checks and balances are unknown. This point can be illustrated by an example, that of the common blackberry. Blackberries were planted for their fruit and to provide suitable food for some of the European birds the early settlers had hoped to acclimatise. They were introduced in Tasmania in 1843; they are now to be found in suitably moist land in much of south-eastern Australia, often on good alluvial soils; they cover some 663,000 hectares of land in Victoria alone (Amor and Harris, 1979). Local control with hormone sprays is now possible, but the work is tedious and expensive, especially in a country where labour costs are high and land values relatively low.

But why are blackberries so invasive in southern Australia? It is instructive to consider how plants such as the blackberry are kept in check in their northern lands, and become rampant in Australia. The answer is that in England, for example, the blackberry functions as a nurse plant, in that it protects young seedlings from premature browsing. In time, the seedling will grow into a tree with a dense canopy that shades out the blackberry, so there is a natural cycle. In Australia, the tree canopy is too open to allow effective shading, and the blackberries therefore grow unchecked to make impenetrable thickets. These thickets are a haven for rabbits, an even more destructive introduction; the blackberries pre-empt the land from productive use or add greatly to the cost of bringing it into productive use – and of course they destroy the indigenous plants of the valleys and banks of the streams, where once there was a much more interesting and diverse range of plants and of animal life supported by them. Control is possible, but eradication is not, because blackberries are spread through inaccessible valleys in the highlands of the east coast, and are easily dispersed from these areas by birds. (Biological control using a rust species has become a prospect in the 1990s, but it raises a familiar problem. Can we be certain that its effects will be limited to the blackberry?)

This example illustrates two points: the first is that they flourish unchecked because the natural biological controls of their environment of origin are absent. The sparse foliage of most Australian trees is an adaptive response to an excess of solar radiation in Australia, whereas the dense canopy of the deciduous trees of countries such as England is a response to the need to maximise photosynthetic activity during a short growing season. The second is that management or control requires continuous, unending human intervention. The lesson implied here is very significant – every such perturbation of natural systems increases the management overload on our species.

A second example is even simpler in its consequences. On the outskirts of Melbourne, a city of nearly three million people, there is an area of rugged country adjoining the Bend of Islands near the Yarra River. It is difficult of access, and characterised by poor soils. For these reasons, it has not been cleared either for agriculture or for suburban subdivision, and it thus carries natural eucalypt forest and the native animals that go with it. Because it is near a large city, however, it has now attracted a special kind of urban fringe dweller, the conservationist. It has been established as a 'conservation living zone'. Those who choose to live there must accept restraints: minimum clearing of the natural bush; unsealed roads; no large gardens of exotic plants (a small kitchen garden is allowed, but no more). Above all, no pets. Cats and dogs are not allowed, and are shot on sight. Since Australian society is still close in many of its attitudes to its English ancestor, this is extra-ordinary, because the English are the world's most dedicated lovers of cats and dogs. But the reason is elementary. Cats and dogs are literally incompatible with the native fauna, especially the birds and smaller marsupial mammals (see colour illustration following p. 174).

This example also serves to make two points: some choices are incompatible. In general, human societies look for a compromise solution to resolve conflict. The compromise in this case might seem to be to allow just a few pets under strict control – one pet per family, taken out for walks only on a lead. But this would not work. Many domestic animals are gentle pets by day, but escape by night to a life of savage hunting. So it is one or the other. The second point is, however, that dogs and cats can only be restricted from limited areas in Australia. They have gone wild or feral over most of the continent, and are extraordinarily widely distributed, even in remote and inhospitable terrain. Eradication is not possible.

These two examples illustrate what can be called 'biological pollution'. It is not a very good term, but it is intended to indicate those forms of environmental degradation that are not caused by the usual industrial and agricultural activities. They are 'biological' in that the problems are created by introduced plants and animals. The problems consist in ecosystem instability and as a consequence of disturbance, a deterioration in vegetative cover over vast areas, accompanied by widespread soil erosion. They constitute 'pollution' in its widest sense of bringing about a decline in environmental quality.

Environmental change in Australia

In Australia on the continental scale, clearing for agriculture and the introduction of sheep, cattle and other animals from the northern hemisphere have had an environmental effect beside which all else is minor. When Europeans entered Australia, they did not come alone. They brought their diseases, which were far more effective in reducing the Aboriginal population than the musket; they brought their livestock, their pets, their cultivated plants, and their weeds and their pests. Among the consequences of this multiple invasion on an isolated ecosystem was acute biological instability. The object has been the same as that of the black settlers, to increase the capacity of the land to sustain human beings. Leaving aside the costs, it must be said that the operation has been brilliantly successful. Australia now produces enough food to feed at least 30 million people, and could certainly produce more. If one accepts that Aboriginal techniques of land management sustained an estimated population of 300,000 to one million, then the human life-sustaining capacities of the continent have been increased a hundred-fold or more since white settlement.

But there have been costs, not all of them widely understood. The sheep and cattle each had four jack-hammers for ripping up the thin skin of soil and vegetation that had known only the gentler limbs of the marsupials, none of them horn-tipped. These differences reflect different adaptations. The hoofed mammals of Eurasia and North America multiplied in the prairie grasslands and open plains, but their early evolution seems to have taken place in the rocky mountains typical of young landscapes. Feet tipped with horn would be well shod on harsh scree and bare rock. But Australia is an old landscape of vast alluvial plains lapping the ruins of worn-out mountains, with soils that are often sandy and soft. The broad pad of the camel is well adapted to soft sand, and the Australian marsupials have comparably broad contact limbs. The newly introduced cattle and sheep imprinted themselves on this old landscape quickly, and their growth was rapid. Before 1788, no hoof had touched Australian soil, but by 1860 there were around twenty million sheep and nearly four million cattle, mostly in its south-eastern quarter; and by 1890 there were more than one hundred million sheep and nearly eight million cattle stocking all but the most arid interior.

The initial impact is dramatically described in a letter written in 1853 by one of the early settlers, John Robertson, to Governor La Trobe. It gives a vivid portrayal of problems evident after only sixteen years of settlement in the Glenelg and Wannon catchment in south-western Victoria:

> When I arrived through the thick forest land from Portland to the edge of the Wannon country, I cannot express the joy I felt at seeing such a splendid country before me . . . The few sheep made little impression on the face of the country for three or four years . . . Many of our herbaceous plants began to disappear from the pasture land . . . the ground is now exposed to the sun . . . the clay hills are slipping in all directions . . . when first I came I knew of but two landslips, now there are hundreds found within the last three years . . . springs of salt water are bursting out in every hollow or watercourse. Strong tussocky grasses die before it, with all others . . . when rain falls it

> runs off the hard ground . . . into the larger creeks and is carrying earth, trees and all before it. Over Wannon country it is now as difficult to ride as if it were fenced. Ruts seven, eight, and ten feet deep are found for miles where two years ago it was covered with tussocky grass like a land marsh. (in Bride, 1969, pp. 167–9)

The smile was wiped off the face of the landscape with remarkable rapidity. 'Wannon country' is more erodible than most, but there are many other areas like it in Australia. Land is valuable in Victoria, so that effective soil conservation measures are under way in the worst affected areas. The problem is much more acute in land of low value, where soil conservation is not economically feasible because of high costs and low returns. This holds for nearly all of arid and sub-arid Australia; two-thirds of the continent is slowly degrading.

The course of events in sub-arid Australia is well illustrated from an area in western New South Wales 125 kilometres east of Broken Hill, now Kinchega National Park. The fossil record is particularly good around the shores of Lake Menindee, and early explorers used it as a base, so its early history and pre-history are exceptionally well documented. Around 18,000–25,000 years ago, there was a diverse fauna including giant kangaroos; *Thylacoleo*, the marsupial lion; *Diprotodon*; Tasmanian devils and thylacine or tigers – fossils indicate that there were at least thirty-nine species of mammals in the area. Twenty-three of these species had become extinct by the time the first Europeans appeared on the scene, but ten new species had colonised the area, giving a total of twenty-six known mammal species. There are only nine mammal species known in Kinchega Park today, so seventeen species have been lost in a hundred years.

The vegetation of the area has also changed greatly (Frith, 1973; Lavery, 1978). The explorer Charles Sturt mentions sand ridges covered with 'pine trees' (probably *Callitris*), hakeas, grevilleas and acacia. Most of these trees have now gone, used by the early settlers for building, firewood, endless fencing, and for stock feed. In times of drought, branches were and are cut from trees and fed to stock. This removal of tree cover increases the risk of erosion, accelerated by overgrazing. Kinchega National Park has been protected from grazing for some years, and there is regeneration of native vegetation, but in much of sub-arid Australia, the overgrazed shrubs are replaced by annual grasses, inedible or unpalatable shrubs, or even bare ground. The topsoil is then blown or washed away, and with it the nutrients on which regenerating plant life would depend. The hard subsoil is not easily penetrated by water, and run-off is increased, leading to further erosion and a silting up of the watercourses and waterholes on which the Aborigines were once able to depend.

Changes in pasture composition have led to marked faunal changes. More grass and fixed watering points for sheep and cattle have benefited the large marsupials, especially the red and grey kangaroo, and in similar country further north, the wallaroo, scarcely mentioned by early explorers such as Burke and Wills, but now relatively abundant. This is misleading if it is taken as an indication that the country is in good health, because the increase in numbers by a few species is offset by a marked decrease in diversity of native species. Nearly all the small mammals have disappeared from country like Kinchega, partly through vegetation change, and partly because of competition with and predation by introduced animals.

Sturt is responsible for the first recorded release of an introduced animal in sub-arid Australia, releasing an old horse at Cooper Creek in 1845. Sixteen years

215

216

Australia Felix: an example, near Canberra, of the pastoral idyll. COLIN TOTTERDELL

later, A. W. Howitt found the horse alive and well while searching for Burke and Wills (Denny, in Lavery, 1978). Wild horses (brumbies) are now found in areas as far apart as the Snowy Mountains and the Kimberleys, although they are only locally abundant. The English wild rabbit was released in 1860 by Thomas Austin at Barwon Park near Geelong in Victoria, and by 1886 there were rabbits in south-west Queensland. Rabbits were spread by men who saw them as sport or food, by rabbiters for whom their fur or carcasses were a living, and by their own talents for colonising a land in which open grassland and sandy soils were common. The devastating effects of rabbits on the drier region of Australia are well documented. Foxes, cats, goats, pigs, donkeys and camels have also run wild (Rolls, 1969). Feral cats are extraordinarily widespread, and very destructive of birds and small marsupials. A recent estimate puts the number of wild donkeys in the Kimberleys at over one million, exceeding the beef cattle with whom they are in competition. The Arabian camel was introduced in 1846 by the Victorian Government, and again in 1860 for the Burke and Wills expedition. It became a common means of transport in inland Australia until it was replaced by the motor vehicle in the 1930s. Large numbers were then set loose, and thousands have since been shot for pet food in central Australia. In recent years, camels have been re-exported (repatriated?) to Arabia. European carp are a serious pest in many Australian rivers, displacing more palatable species and fouling the water. Some of these introductions have not been especially destructive, although they have tended to replace native species, much as

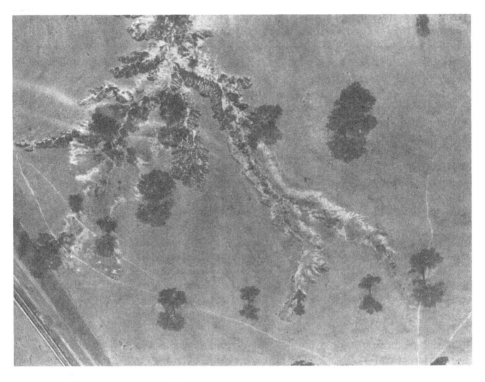

Australia Infelix: an example, near Canberra, of the pastoral reality in too much of Australia. This area is now under control, but control is more the exception than the rule. FRANK INGWERSEN

the dingo replaced the thylacine, and as the house sparrow has replaced the colourful zebra finch in many outback settlements. This is a loss in variety, but minor compared with the destructive effects of the rabbit, wild pigs, and water buffaloes in the Northern Territory. The following table presents comparisons of mammals recorded 125 years ago and their status today.

Many plants have also been extraordinarily destructive in Australia. The prickly pear (*Opuntia*) in Queensland, the noogoora burr in the dry plains, and a host of other 'weed' plants have taken hold wherever conditions favour them. There are Mediterranean docks and sorrels on the Bogong High Plains; lantana chokes disturbed rainforest; Chinese privet (*Ligustrum lucidum*) forms a dense understorey in the bushlands of North Sydney, with dying *Angophora* above them; veld grass (*Ehrharta* spp) from South Africa smothers the incomparably beautiful ground plants in Kings Park in Perth, while *Phytophthora cinnamomi*, a plant pathogen, infects the roots of jarrah, banksia, and much else.

Management and its costs

There has been a great deal of research into pest control in Australia, and there are many success stories. The control of prickly pear is an outstanding example. Several species of *Opuntia*, a large, spiny, cactus-like plant, were introduced to Australia for various purposes – for hedges, stock feed, fruit or as garden plants. Prickly pear

Mammals Recorded by the Blandowski Expedition to the Region of the Murray–Darling Junction in 1856–57 and Their Status in that Region Today

Species		Present Status
Platypus	*Ornithorhynchus anatinus*	Common
Echidna	*Tachyglossus aculeatus*	Common
Yellow-footed Marsupial Mouse	*Antechinus flavipes*	Extinct
Brush-tailed Phascogale	*Phascogale tapoatafa*	Extinct
Red-tailed Phascogale	*P. calura*	Extinct
Fat-tailed Marsupial Mouse	*Sminthopsis crassicaudata*	Common
Common Sminthopsis	*S. murina*	Rare
Antechinomys	*Antechinomys laniger*	Extinct
Western Native Cat	*Dasyurus geoffroii*	Extinct
Numbat	*Myrmecobius fasciatus*	Extinct
Short-nosed Bandicoot	*Isoodon obesulus*	Extinct
Pig-footed Bandicoot	*Chaeropus ecaudatus*	Extinct
Barred Bandicoot	*Perameles bougainville*	Extinct
Rabbit-eared Bandicoot	*Macrotis lagotis*	Extinct
Ringtail Possum	*Pseudocheirus peregrinus*	Common
Brush-tail Rat-kangaroo	*Bettongia penicillata*	Extinct
Lesueur's Rat-kangaroo	*B. lesueur*	Extinct
Rufous Rat-kangaroo	*Aepyprymnus rufescens*	Extinct
Brown Hare-wallaby	*Lagorchestes leporides*	Extinct
Bridled Nail-tail Wallaby	*Onychogalea fraenata*	Extinct
Crescent Nail-tail Wallaby	*O. lunata*	Extinct
Red Kangaroo	*Megaleia rufa*	Common
Grey Kangaroo	*Macropus* spp	Common
Eastern Water-rat	*Hydromys chrysogaster*	Common
Gould's Mouse	*Pseudomys gouldii*	Extinct
Sandy Inland Mouse	*P. hermannsburgensis*	Extinct
Brown Desert Mouse	*P. desertor*	Extinct
White-tipped Stick-nest Rat	*Leporillus apicalis*	Extinct
Stick-nest Rat	*L. conditor*	Extinct
Mitchell's Hopping Mouse	*Notomys mitchellii*	Very uncommon
Dingo	*Canis familiaris dingo*	Extinct

(Source: Frith, 1973)

quickly became naturalised in northern New South Wales and south-eastern Queensland. By 1925, about twelve million hectares of land was rendered completely useless by prickly pear, and another twelve million was severely infested. Control had been attempted through spraying, but had not been effective. After extensive biological research, the eggs of a moth, *Cactoblastis cactorum*, were introduced from Argentina, and achieved spectacular success in the warm dry climate of the infested areas of south-east Queensland. In the cooler and wetter tablelands of northern New South Wales, control with hormonal sprays has been slower. It is noteworthy that *Cactoblastis* has no unwanted effects, in that its feeding and breeding are restricted solely to cactus-like plants. Many earlier attempts at biological intervention in Australia produced serious consequences other than intended ones, but such mistakes are now guarded against.

A recent example of successful biological control is the introduction of a South American beetle, *Cyrtobagous singularis*, to eat a pondweed, *Salvinia molesta*, that has covered the surface of lakes, rivers, dams and irrigation canals in much of Australia. It blocks ditches, drains and pumps, so is especially troublesome in irrigation areas and water storages, but it also interferes with fisheries and crowds out natural plants. The beetle was carefully studied before release to guard against unwanted effects. In areas in which it has been released it has now achieved a biological balance; most of the weed has been cleared, and there remains now only small areas of *Salvinia* and small populations of the beetle.

The best-known and most spectacular example of biological control in Australia – perhaps in the world – was the introduction of myxomatosis to populations of wild rabbits in 1950. This is a viral disease spread by mosquitoes, and its early success was quite spectacular. By 1953, most of Australia was virtually free of rabbits, with a kill estimated at around 90 per cent. One result was that the annual wool clip increased suddenly by 30 million kilograms. The rabbit has been almost unimaginably destructive in Australia, and has played the leading role above all else in degrading the environment. The early success with myxomatosis should have been followed up with a vigorous campaign using conventional means such as fumigating and ripping the warrens and shooting and dogging the survivors. This golden opportunity was not taken, and rabbit populations have built up again in some areas, although not in the plague proportions that preceded the introduction of myxomatosis (Rolls, 1969).

Biological pollution in Australia, in the sense that the term is used here, has had two prime costs. The first is the loss of, or decrease in, a rich indigenous fauna and flora in many areas. This loss is aesthetic, scientific and perhaps also spiritual, but it is not total, and some loss was inevitable given colonisation of such a land by Europeans. It is still possible to have a system of reserves to conserve flora and fauna, and although they require highly skilled management to serve their function, this is within our national resources.

The second cost is even more serious. The deliberate and accidental introduction of pests, weeds and diseases from overseas has significantly reduced the productivity of the land in many ways, but above all through a reduction in plant cover and in accelerated soil erosion. Thus the renewability of what we optimistically call 'the renewable resources' is made more difficult. Maintaining the integrity of 'renewable resources' is a far more serious problem in Australia than the problem of industrial pollution; this may also be true world-wide.

A recent national report on Soil Conservation in Australia (Pauli, 1978) concludes that 51 per cent of the total area used for agricultural and pastoral purposes in Australia was assessed as needing some form of soil conservation, and that remedial measures required to maintain present levels of productivity (*not*, let it be noted, to restore the land fully) were $A675 million at June 1975 prices. But that is not the key problem. The necessary steps probably will be taken in non-arid agricultural Australia, because it is economically justified. The critical, almost insurmountable, problem is that of arid and sub-arid Australia, which constitute two-thirds of the continent.

The arid grazing lands generally have the lowest priority in terms of value of

219

production and fixed assets relative to the cost of soil conservation works. As the measures required for treatment in the degraded arid areas typically include some destocking measures, the implementation of conservation programs is likely to result in even lower returns – at least in the short term. Where action is most urgent, the arid areas where rehabilitation is difficult and slow, the economic return is low. The dilemma that this presents has parallels in most 'third world' countries, where remedial measures are essential to the continuing survival of the biosphere, yet money invested will not show an immediate economic return.

Conclusion

Island ecosystems that are in steady communication with the rest of the world are bound to suffer disruption. The world as a whole moves towards a global ecosystem. Specific areas can be managed for specific purposes, but self-regulating systems are inescapably being replaced by managed ones.

The experience of Australia highlights these problems. The productivity of what are called the 'renewable resources' is a function of ecosystem management, and this should be a major theme. A special problem is created where technical priorities for action – as in the case of soil conservation in most of northern and Central Australia – do not coincide with economic priorities. In Australia, one might conclude that a proper use of part of the profits from non-renewable resources (mining) should be put to the rehabilitation of degraded ecosystems.

Postscript

Our species has shown remarkable management skills through time, but there is the possibility that we are incurring management overload, with the increasing prospect of human error. I explored this further in 'Managing a Resources Boom: Or In Praise of Country Boys' (Seddon, 1982).

220

References

Amor, R. L., and Harris, R. B., 1979, 'Survey of the Distribution of the Blackberry (*Rubus fruiticosus* L. *aggr.*) in Victoria and the Use and Effectiveness of Chemical Control Measures', *Journal of the Australian Institute of Agricultural Science*, vol. 45, pp. 260–3.

Bride, T. F. (ed.), (1898) 1969, *Letters from Victorian Pioneers*, new edn ed. C. E. Sayers, Melbourne: Heinemann. First published for the Trustees of the Public Library, Museums, and National Gallery of Victoria by Robt. S. Brain, Government Printer, Melbourne.

Frith, H. J., 1973, *Wildlife Conservation*, Sydney: Angus & Robertson.

Turner, J. S., 1966, 'The Decline of the Plants', in A. J. Marshall (ed.), *The Great Extermination: A Guide to Anglo-Australian Cupidity, Wickedness and Waste*, London: Heinemann.

Lavery, H. J. (ed.), 1978, *Exploration North: Australia's Wildlife from Desert to Reef*, Richmond, Victoria: Richmond Hill Press.

Rolls, E. C., 1969, *They All Ran Wild: The Story of Pests on the Land in Australia*, Sydney: Angus & Robertson.

Pauli, H. W. (ed.), 1978, *A Basis for Soil Conservation Policy in Australia*, Canberra: Australian Government Publishing Service.

Seddon, G., 1982, 'Managing a Resources Boom: Or In Praise of Country Boys', in S. Murray-Smith (ed.), *Melbourne Studies in Education*; Melbourne University Press.

23 The lie of the land:
Australia all over

STOCK MEN

Winter in Perth, and the Claremont cowboys,
The Joondalup Jackeroos, are out again,
Heading north in their Nissan Patrols
Beyond Daydream and Cue, beyond Youenmi,
Cutting up the Tanami Track
Moleskinned from R. M. Williams in City Arcade.
Thoroughly outfitted,
They ride the range in their Rangeriders.
Unmoved by Black Deaths in Custody
(Not my problem, mate)
Nor a brutal convict past
(I wasn't there, was I)
Lured, perhaps, by a desert mirage of She,
Now Mrs Paul Hogan the Second
And no longer inviting rescue,
They lacerate the thin skin of the desert,
A casual cat-o'-nine-tails across its bare back.

This second violation of the land
Worse than the first, because so trivial.

My mental map

Australia is inside our heads, your Australia in yours, my Australia in mine; they have been put there by what we have read, learned, and seen, and insidiously, by the media and their image-makers, to which we all succumb in part and resist in part. As continents go, Australia is mostly low, flat, and dry, with landscapes that are old and soils poor in nutrients. That much is fact. But to say that it is drought-ridden, or that the soils are infertile, is to go beyond fact. Drought has no physical reality. The physical reality is that of markedly variable rainfall. Drought is a problem of perception, applied when Nature has failed to meet our expectations. We find it hard to modify these expectations, but Nature is not under contract to us. She will not adapt to our ways: we must adapt to hers. Above all, we must pay close attention to her tolerances, which are sometimes surprisingly broad, and sometimes narrow. In Australia, it is generally the latter.

In showing you the Australia that I see, there are some facts, some speculations, some images, some media images, and some questions. First, where are we? At a dinner party last year, a rather snotty English doctor now working in Perth made some disparaging remarks about 'the Antipodes'. Feigning surprise, I said 'But I didn't know that you were from Bermuda: I thought you were from London'. Enraged incomprehension. Then I had to explain that if we draw a line from Perth through the earth's centre, where it comes out the other end – at the antipodes – is very close to Bermuda, Prospero's isle; if, on the other hand, we draw a line from London through the centre, it comes out, appropriately, at a small, cold, wet, uninhabited island, named 'Antipodes' by Cook, who knew his latitudes and longitudes. This island is some six hundred kilometres south-east of Dunedin in New Zealand, or as far south again from Hobart as Hobart is from Melbourne. If we were indeed antipodean to England, we would be better watered. Western Australia especially runs out of 'south' all too soon; lying between 10° and 43° of latitude south, Australia has its widest extent in the thirties, the zone of maximum climatic uncertainty. It is the only southern continent to do so – Africa has its widest extent in the humid tropics, although it repeats Australia's configuration in the northern hemisphere.

Australian climates range from that of Innisfail to that of Alice Springs and are therefore hardly uniform, but the 'Australian Climate' has nevertheless been extravagantly praised. Most of us like the sun, and find pervasively grey skies depressing. Perth people are so proud of their climate, you would think they had invented it, despite the tearing winds of spring and the blistering forty-degree temperatures in summer, when the sun is the enemy. In the nineteenth century, Australia was heavily promoted as a health resort, especially for consumptives. That leading British medical journal, *The Lancet*, carried several letters in the 1830s reporting on the absence of the epidemic diseases in New South Wales, and another letter warning the British doctor against going there, for without a return ticket, 'he must become a clerk or a cattle driver; or he must starve', since there would be no call for his professional skills (*The Lancet*, 1, 1837–38, pp. 716–17). However, there were also some more cautious reports: the Royal College of Physicians, for example, claimed that consumption in Australia was actually more common 'than from the mildness of the climate might be expected, and more in advanced life suffer from the disease than in England' (Powell, 1978, p. 131). Medical personnel today rarely form a part of the unemployed in Australia.

The notion promoted by the media of Australia as a 'Big Country' is one to which I have a mixed reaction. I prefer to think of it as a small country with big distances: the 'tyranny of distance', Geoffrey Blainey's fine phrase, is still with us, and transport costs are high because of it. The sense of distance is dependent on perceived travel times, however – a fact brought home to me some thirty years ago when talking in Perth to a newly arrived professor of history, whose field was English medieval history. 'Why', I asked, 'would someone with your interests want to come to Western Australia?'. 'Because', he replied, 'Britain in the Middle Ages and Western Australia in the 1960s have a great deal in common'. He pointed out that both were dependent on a mix of an agricultural and a mining economy. Population densities were low. Although they were much higher in Britain, they were comparable if you measure distance in travel-time. Britain and Western Australia were much the same size measured in travel-time. In fact Britain was bigger.

I suppose he was right. The 'Big Country' rhetoric irritates me because it so easily leads into the cornucopian view of Australia, that there is an abundance of resources, plenty more where that came from – a view conspicuous in our National Anthem. This is not an easy country (if I may say that, after my discussion of the pathetic fallacy in Chapter 2), and it never will be. Agricultural Australia is a small country: our wheat production for example is close to that of the United Kingdom and sometimes drops below it; the average Australian is incredulous on being told that in 1972, for example, the United Kingdom grew one and a half times as much cereal as Australia, on one-third the area. That was a bad year here and a good year there; moreover, both acreage and yields have increased substantially in Australia. But the fact remains that we are not well endowed in life-supporting renewable resources at the continental scale. It is a large country only as a communication system – the tyranny of distance still applies. Australia is a small country with long journeys. Much of our wealth, the wealth that pays for our standard of living, is non-renewable. We produce less of almost every agricultural product, except wool, than France, and even Italy, one-third the size of New South Wales and much of it mountainous, has a higher total cereal production (counting maize, rice and barley) than we do. We are a substantial exporter of cereal, but not a major producer.

But the distances remain huge. London to Istanbul or Moscow is less than Perth to Sydney, and the flight across the continent is wearisome to those who make it often. It takes its toll on politicians. Travel by road and rail is slower, for the obvious reason, but also because it is more circuitous. In giving a thumbnail sketch of Australia to Americans, I sometimes ask them to superimpose the two countries, which are of comparable size. Then I say, if you are driving from New York to Los Angeles, but across Australia, you would find yourself in Nevada before you got through Indiana and you would stay in Nevada until you were almost in Los Angeles. No rich mid-West, no Mississippi or Missouri, no Rocky Mountains, not even the productive Central Valley of California – and above all, *there would be no road* unless you detoured well to the south. The last is what really hits Americans, since they have a choice of routes across the country at every latitude, while our road skirts the coast, as does our only transcontinental railroad. It is not so much a big country as an empty one, although not absolutely: there is not much 'desert' in either of its two literal senses: without vegetation or without people. Although there is some sandy desert and some stony desert, even that generally has some plant cover. It is not like the Sahara. And most of it is occupied by Aborigines and pastoralists, although at very low densities.

Popular images

One of the most persistent images of Australia is that of its antiquity: 'this old, old continent, the land where time stood still'. Australia, or parts of it, are old in some highly significant senses, and much of the distinctive character of the continent follows from this (see colour illustration following p. 174). While endorsing the idea, I want at the same time to question what the media do with it. Behind the factual claim, there is often an implied boast and an implied excuse, sometimes coexisting uneasily. With a country so immeasurably old, with Uluru (Ayers Rock), who needs Gothic cathedrals or the Colosseum, comparatively recent events? And in a land that is worn out, exhausted with its millennia of being, it is unreasonable to expect too much of its people. American notions of efficiency do not sit well in a tired land.

One needs some knowledge of earth history to comprehend Australia. Most of our soils are old, and some of the landscapes have had much their present form for a very long time, the reason being that Australia is relatively stable tectonically compared at least with Japan or New Zealand, or the Andes, California or Italy. There are huge fault systems along both the south-eastern and south-western coast-lines: the Darling Fault bordering the Swan Coastal Plain has a throw of many thousands of metres and is one of the world's major lines of movement in the past. But there have been no great mountain-building episodes since the Palaeozoic, and long erosion has made this the flattest of the continents. It also had an uneventful Pleistocene compared with almost everywhere else. It has long seemed to me that when the continents are being handed out, the question one should press hard is 'Did you have a good Pleistocene?', because glaciers and continental ice-sheets are among the great soil-makers, grinding up fresh rock full of the nutrients most plants need. The Pleistocene gave America the Great Plains, China its fertile loess, and much of central Europe its good soils. Glaciers, with their deep V-shaped valleys, also provide good dam sites and, when they meet the sea, deep harbours.

The other great soil-makers are rivers and volcanoes: the fertile soils of Java are volcanic; the valleys of the Po in Italy, the Nile, Tigris, Euphrates, Danube, Rhine and Rhone all bear witness to the soil-making capacity of the world's great rivers. We have no great rivers on that scale today, and although we had them as late as the Pliocene, their deltaic deposits are drowned by the rise in sea-level. We have no active volcanoes. Cenozoic volcanics in eastern Australia have yielded some good soils, but on the continental scale, most of our soils are old and leached of nutrients, especially low in phosphorus and trace elements. Reflecting on these elementary facts of place has led me to a 'linking' assertion, attempting to clarify human history by underwriting it with earth history: namely, that all the old civilisations come from new lands, flourishing on soils fresh-made, from volcanoes, or great rivers, or glaciers. Nutrient-rich soils create a high-energy environment in temperate lands (in the tropics, heavy rainfall and rapid oxidation of organic matter tend to deplete the nutrient store).

The high-energy environment left by the Pleistocene has had a further consequence in Europe, which differs from North America and Asia in the alignment of its mountain chains. In North America the two major chains run north–south, so that fauna and flora could migrate south ahead of the advancing glaciers and return as they retreated. Asia and China have some major east–west chains, but they do not form almost continuous barriers like the Alps, Pyrenees, Maritime Alps, Dolomites and Carpathians. These had the effect of blocking movement, especially of plants, so Europe has an impoverished flora compared with China and North America. At the end of the Ice Age, Europe was like an empty race-track waiting for the starter's gun, or better, perhaps, like a major department store at the moment the doors open for the first day of the New Year sales. The race goes to those who let nothing stand in their way. As Tim Flannery has shown so well in *The Future Eaters* (discussed at greater length in Chapter 24), rapid recolonisation of Europe as the glaciers retreated encouraged what is essentially a weed flora and fauna, made up of aggressive, opportunistic species with high energy demands and the strongly competitive traits which enabled rapid expansion into every available niche: which is to say, the driving characteristics of world history, and perhaps, thus, one of the primal causes of both World Wars. When the aggressive colonisers of Europe had run out of new worlds to conquer, all that remained was to overrun one another.

Despite the above, Australia is not the oldest, but the youngest, continent, along with Antarctica, from which it split in the Eocene. The notion of Australia as the 'Timeless Land' makes no sense. Both physical and biological change have been operative here as elsewhere. Physically, even the last 20,000 years have seen massive changes in some areas, such as the great outpouring of basalt in western Victoria, and a rise in sea-level which broke the links between the mainland and Tasmania to the South and New Guinea to the north. The effects of rising sea-level are displayed visually in the rock-shelter paintings in Kakadu at Nourlangie Rock, Ubirr and other sites. They show that the environment some 30,000 years ago was part of an inland plain some 300 kilometres from the sea to the north. About 8,000 to 10,000 years ago, the sea-level rose, rainfall increased, low woodland became replaced by forest in some localities, and salty estuarine conditions prevailed some distance inland, replaced in turn by fresh-water conditions around 1500 years ago. There was then a great increase in waterbirds like the magpie geese, which now sometimes blacken the sky. Each of these changes is reflected in the food sources depicted in the rock paintings: for example, dolphins are shown in the salt-water estuarine period, but not in the pre-estuarine or post-estuarine periods. There is even a thylacine of the pre-estuarine period at Ubirr, long extinct in this area.

The 'Dead Heart' image is an extension of the 'Timeless Land' image. If time stands still, it is because the heart of the land has stopped beating. But Uluru is about as dead as any other rock, and the Japanese tourists would be more than surprised if it were to begin pulsating in that clear air. The view of Australia as the land that time forgot began early, and was entwined with antipodean fancies, where the natural world was stood on its head, where swans were black rather than white, and all was paradox. The initial response to the Australian flora and fauna can be forgiven, but it has persisted too long, as has the view that Australia was a refuge in which relatively primitive animals persisted long after they had been wiped out by tough competition elsewhere. The present Australian fauna is not ancient, but of recent origin, and it is highly adapted.

The flatness of the continent exaggerates the effects of aridity. The arid south-west of the United States is crossed by big rivers rising in the Rockies and Sierra Nevada, whereas the Murray-Darling only just makes it to the coast, while rivers to the north of the Darling do not. The western slopes of the Great Dividing Range provide an ample catchment for a substantial river, and Coopers Creek and the Diamantina are often well supplied with water, although there is great variability. I celebrated my sixtieth birthday by canoeing with a group of friends down Coopers Creek from south-western Queensland to the last camp of the only survivor of the Burke and Wills expedition, at Quiapidrie below Innamincka. It was here that A.W. Howitt's party found King by one of the deep, permanent waterholes that mark the course of the Cooper and its distributary, Strzelecki Creek. We were able to paddle continuously in April, and were surprised by the amplitude of the river, and above all by the grandeur of the river red gums. George McGillivray, who travelled down the

Diamantina to the Cooper in 1870, reported the following year in the *Australasian*, 'I crossed no country anything like it since I crossed the dividing range from the Gulf of Carpentaria. The gum trees are the finest I have ever seen'. The word 'waterhole' understates the quality of these great sheets of water, at Nappamerie or Bulloo Bulloo, many kilometres long, up to ten metres deep, wide as the Murray at Echuca, and never dry. This is no creek: although subject to drought and flood, it is a river – but, as A. D. Hope tells us, it is 'drawn among inland sands'. Lake Eyre is below sea-level, and these rivers fail to reach the coast, not because they run out of water, although that, too, can happen, but because they run out of slope, breaking up into a maze of distributary channels and shallow lakes. Australia, along with the Netherlands, is one of the low countries, or, as the physiographers say, it lacks relief.

So Australia is low, flat, dry, and fire-prone, with landscapes that are old and soils that are exceptionally low in nutrients, having passed through countless weathering cycles without renewal. Yet this is also part of the reason why there is such biodiversity. Europe is ecologically depauperate; the fauna and flora could not migrate south ahead of the continental ice-sheets, and return north in the next interglacial, their movement blocked by the east–west mountain barriers. The consequence is that Europe is still in the process of recolonisation after the last Ice Age, with a simplified biota. It was the very simplicity of the relationships that enabled the science of ecology to be born in Europe.

Yet the familiarity with fertile soils, good and reliable rainfall and a simple biota gave European colonists a very poor background experience for understanding and managing the Australian environment. The biases of Eurocentric thinking have coloured and distorted the perceptions of scientists as well as non-scientists, and we are still not free from such distortions, partly because we are still struggling to understand Australia to this day.

The physical character and the evolutionary history of the continent between them are the two major biological determinants. In breaking off from Antarctica about 40 million years ago, Australia completed the break-up of the ancient Gondwanaland: India and Africa and South America had abandoned us earlier. It was not until about 15 million years ago that our continued northwards movement brought us into collision with the Asian plate, marked by the line of the Markham Valley in New Guinea; Asian plate to the north of it, Australian plate to the south. Thus for 30 million years we were completely isolated, and we are still separated by a shallow sea from Papua New Guinea. This long isolation has made us exceptionally vulnerable to invasion by introduced species.

The fauna and flora

The unique biota of Australia (see Braithwaite, 1990a) is usually divided into six elements. The oldest goes back beyond Gondwana to the super-continent of Pangaea; representatives are the cycads, the cockroaches and the dipleurans. Some of the creatures with very ancient lineages were noted earlier in the discussion of the caves of Cape Range in north-western Australia (see Chapter 9). The next is the Gondwanan biota, which shows marked affinities with the other Gondwanan lands, notably the Proteaceae from southern Africa, the southern beeches (*Nothofagus*) and the gymnosperms, such as *Podocarpus*, or *Araucaria* (this last is found in New Caledonia and Chile), and animals like the large flightless birds, the parrots, some geckos and some turtles. Next there are the many species that evolved in Australia during its period of total isolation, such as the eucalypts and the honey-eaters. As we collided with the Asian plate, some degree of contact was renewed; most of the snakes and lizards, and many birds, rodents and insects came in during this interval in the Tertiary. The fifth element comes from human contact: for example, the Aboriginal hunting dog, the dingo, arrived only 3000 to 4000 years ago. Native rats and the rainbow bee-eater (*Merops ornatus*, a summer visitor in our garden) also belong to this period.

Finally, the European invasion of the last 200 years has brought a substantial range of introduced plant and animal species, notably the horse, the sheep and goat. Soil compaction and erosion were the almost immediate outcome. Loss of habitat and European-induced environmental change also led to rapid extinctions. Even in sparsely populated Central Australia, about 40 per cent of mammal species has disappeared in the last 200 years. This process of extinction is rapidly reducing bio-diversity, since the successful, introduced species, like the rabbit, are generalised rather than specialised, and thus a few species can cover a wide range of habitat. However, the process of extinction goes back beyond the European invasion; a major biological event of the late Pleistocene has been the extinction of the megafauna, our large mammals. This extinction post-dates the arrival of our species, and some biologists link the two events, although climatic change towards increasing aridity has clearly also played a part. Perhaps the question should not be why did Australia's megafauna die out, but why has the megafauna survived in Africa and Asia? A part answer may be that it has survived where it has evolved *along with* the hominids.

The interaction of evolutionary history with the physical environment

Next, we need to see how the physical state and our evolutionary history interact. Perhaps the most dramatic difference from Africa is that Australia's is primarily an insect-driven ecology. Invertebrates are often prominent in an infertile landscape and in large areas of Australia the primary herbivores are invertebrates. In place of springboks and antelopes, we have termites and grasshoppers. Termites can break down cellulose with very low nutrient content, they can fix nitrogen using symbiotic micro-organisms in their hindgut, and they store grass in the termitaria and are thus able to survive harsh dry periods. Grasshoppers take the other tack – they are opportunists, with a very high rate of increase during favourable periods. The invertebrate herbivores provide abundant food for insectivorous vertebrates; thus the arid zone of Australia supports the highest diversity of lizards in the world (see colour illustration following p. 174). We are also rich in seed- and insect-eating birds – all those finches and budgerigars, for example. The mammalian vertebrates are mostly very small, and often nocturnal, to avoid the heat and desiccation of the day. Their predators, like the dasyurid 'native cats' (these are marsupials, not true cats) are also nocturnal and also small. For this reason alone, the majority of Australians never see most of our mammals in action in the wild.

In 1987–88 I was invited to make an environmental impact assessment for a proposal to run an electric power line in the Northern Territory from Marrakai to Jabiru, thus linking Kakadu to Darwin. While I was there, three experiences particularly impressed themselves on me.

The first was that a young CSIRO scientist, Alan Andersen, showed me his ant patch, in which there were not just many ant colonies, but many species of ant.

Many of the species were seed-eaters, and many of them ate the seed of only one or a very few species of grass. Alan could identify the different species from the seed husks that fringed the entry to each nest. Other species, however, were meat-eaters, with a more varied diet. Some of them ate the seed-eaters. Here, in a small space, was a whole world with a complex ecology, including a full range of predator–prey relationships, but a world only visible to those who had learned to see it.

The second memorable experience was that Tony Press, the research biologist at the national park headquarters, took me with him in the very early morning to check his traps, and I thus saw a range of small marsupials I would never otherwise have seen. Little insectivores, *Antechinus bellis*, were the most common animals trapped, and they were carefully weighed, checked, tagged and released. A few times there was a native cat or northern quoll, *Dasyurus hallucatus*, a fierce carnivore. The picture that I began to build is that Australia is as rich in herbivores as Africa, but that in place of the antelope, there are insects. If a herbivore, from an ecological perspective, is the primary digester of cellulose, then this role is filled in much of Australia by insects, to which the large termitaria across northern Australia bear witness, but then so do the 'white ants' that aim to break down the timber of our homes further south. Especially in northern Australia, our ecology is insect-driven. Australia has about 1000 species of social insect, while the USA has 400 and the UK, 46. There are 182 species of termite alone. Europe has 2 (data from Groombridge, 1992, pp. 110, 112).

Let's hear it for the termites

Another interesting mutualistic relationship is that between termites and eucalypts: the termites eat parts of the inner trunk, which is of course no longer living tissue, but represents nutrients otherwise locked away from use; the termites gain food and protection, while returning nutrients to the tree in usable form. They do far more than this – in fact they make an immense contribution to the ecology of the wet–dry tropics. 'They are not the inert tombstones they appear to be', says Braithwaite, (1990b) (see colour illustration following p. 174). He explains that in creating hollows in trees, they make homes for one-fifth of the bird and reptile species in 'the Top End', half of the mammals and a quarter of the reptiles, all of which contribute nutrients to the tree from their droppings. The termitaria themselves also provide shelter to huge numbers of invertebrate species, geckos, lizards, snakes, and as the mounds decay, the hooded parrot, kingfishers, marsupial mice, the native cat, goannas and bandicoots. Their constant temperature helps reptiles to remain active and assists in the incubation of eggs. They offer refuge during floods and fire. Above all, the termites break down cellulose ('They are stationary cows. They are a vacuum cleaner removing the natural debris of the system') while themselves constituting a year-round larder for insectivores (Braithwaite, 1990b, pp. 306ff).

Termites have played an indirect part in the evolutionary biology of Australia. For example, ornithologists from North America and Africa note the absence of woodpeckers in Australia. Why are they absent? Because their skill has not been needed. The abundant nesting holes made by the termites are much used; thus 21 per cent of our non-passerine birds are hole-nesters against 7 per cent in southern Africa – but we have no species which make their own holes, like the woodpeckers.

Scarce and variable food supplies put a premium on mobility, especially for birds; for example, 26 per cent of Australian birds have been classified as nomads, very high by world standards, while only 8 per cent are north–south migrants. A few of the larger mammals, such as the red kangaroo, are also able to track resources over a large area, as of course were the Australian Aborigines, most of whom were semi-nomadic, although they moved around within fairly well-defined territories.

Plants are not nomads, except in the special sense that some plants, particularly ephemerals of the desert, have short life spans, widespread seed dispersal and rapid germination and growth in favourable conditions, so that the species moves over large areas, even if the individual does not. Perennial plants must be very highly adapted to local conditions, which almost always include low-nutrient soils and variable rainfall. Thus about 300 species of legumes are capable of fixing nitrogen through nodulation. So can many species of *Casuarina*, and mycorrhizae (associations between roots and fungi that can assist nutrient uptake) are also common, especially among the eucalypts. Such symbiotic and mutualistic associations are characteristic of a low-nutrient environment, but not of our imported food crops, which demand heavy inputs of fertiliser.

Birth of the eucalypts?

The third memorable experience at Kakadu was something of which I was already dimly aware, but had never grasped fully. Some northern eucalypts are deciduous, with new leaves that are bright green, and some have a good canopy, all contrary to our southern view of the eucalypts as being hard-leaved grey-green evergreens with a scant canopy. This experience combined with another I had had in Kalimantan Timur (once East Borneo) a few years earlier when I was conducting a review for ADAB (the Australian Development Assistance Bureau) of our aid programs to the University of Mulawarman in Samarinda. Here I was able to visit a patch of 'heidewald' (heath-forest), so called by Winkler, a German botanist who described it, within the more typical rainforest of Kalimantan Timur. (The major research on heidewald is by Brunig, e.g. Brunig, 1966; there is a review of the formation in Whitmore, 1975.) The rainfall here was the same as elsewhere, but the plants were nevertheless often subjected to water-stress, because they were on a pod of deep, rapidly draining sand. The species were generally closely related to the luxuriant rainforest nearby, but one could almost have been in Australia. The Myrtaceae were much in evidence, as they are in the rainforest. But these trees were sclerophylls, with thickened leaves. They were more grey than green, reflecting the desiccating sun better.

Even more striking was the litter. These siliceous sands are very deficient in nutrients. Nearly all available nutrients are locked up in the plants themselves. But the bark of a tree is dead, and nutrients locked up in bark are taken out of circulation. Not, however, with these trees. The bark peeled off in strips, mulched the soil, and eventually broke down and became available for recycling. Moreover, this dry litter was fire inducing; even in the wet tropics there can be dry spells of a few weeks, and lightning is common enough. These sclerophyll patches burned, and had done so for a long time. The plants had become fire-adapted, and in some cases actually required fire to split open woody seed-cases onto the new ash-bed, briefly rich in nutrients, assisting the first, rapid flush of new growth after rain. All of this is very familiar to any observant Australian, but I was seeing a fire-adapted, litter-creating, sclerophyll vegetation evolving in the midst of tropical rainforest, and thus witnessing what comes close to a re-run of evolutionary history in much of Australia, for our present sclerophyll flora is of recent origin.

Another feature of the eucalypts is that their leaves are subject to an exceptionally high level of insect attack – between 12 and 40 per cent of leaves are insect damaged, and this too has an ecological feedback. The leaf litter from eucalypts breaks down very slowly through simple weathering to release nutrients, but rapidly in passing through the gut of an insect.

The flora and fauna are linked in a multitude of ways, since they are co-adaptive. Most plant dispersal is by invertebrates, and seed dispersal by ants is common. Another form of plant dispersal is by very small wind-dispersed seed protected in hard, woody capsules, and this is so common that it has led to the spectacular radiation of Australian parrots. One-sixth of the world's parrots, including nearly all the cockatoos, are Australian, feeding on the seed from species of *Eucalyptus, Casuarina, Hakea, Grevillea, Banksia* and *Acacia.*

Another close link between flora and fauna is that Australian woody plants produce copious quantities of nectar, so we have an abundance of nectar-eating birds: in fact Australia and New Guinea have most of the 159 honeyeaters of the family Meliphagidae. Nectar is a significant food source for many birds even in southern Australia (up to forty species) whereas there are none in Europe or North Africa, and only a handful in the other continents at equivalent latitudes, except for southern Africa, which shares with Australia a flora in which Proteaceae and Myrtaceae are prominent. Hence the lorikeets and wattlers and the singing honey-eaters and little brown honeyeaters that do acrobatics in our garden in Fremantle, where we are lucky enough to have our own birds, not yet ousted by blackbirds and starlings, Ceylon crows and sparrows, the winged vermin of most of the world.

Sunlight is rarely a limiting factor in either southern Africa or Australia, so sugar is cheap for the plant to produce. Despite low nutrient availability, energy is easily manufactured from such abundant solar energy, and this is also the probable reason that many Australian trees can tolerate such high loads of sap-sucking insects. These in turn attract ants, which seem to act as a deterrent to browsing herbivores, but in turn attract ant-eating predators, including birds.

Linkages of this kind are typical of our food-chains, and I could expand the story in many ways, for example by talking about reproductive strategies, but this is no more than a sketch, and sketches are sketchy. I have tried to show, by giving a few examples, the interdependence of a rich biota, with a series of relationships between flora and fauna, and between both and the physical environment. There are many more, including a whole series of adaptations to fire. Not only the plants, but also some of the mammals, are fire-adapted. For example, the banded numbat, a beautiful marsupial herbivore of south-western Australia, needs mature forest for breeding and protection, but freshly burned forest for 'green-pick' browsing. Thus only a mosaic suits its needs, a common outcome of Aboriginal patch-burning. Unless its special needs are catered for, however, the forests either do not burn, which starves the banded numbat, or erupt in fierce wildfire, which roasts it.

These examples are enough to give some idea of Australia as a functioning ecosystem; although I make no claim to be an ecologist myself, I have been lucky to see much and to listen attentively in the course of a long life. Australia seen thus is a place to which monocultures are, in general, singularly ill-adapted. It is also one to which men and women from the temperate lands are also poorly adapted, although we are learning to change.

References

Braithwaite, R., 1990a, 'Australia's Unique Biota: Implications for Ecological Processes', *Journal of Biogeography*, no. 17, pp. 347–54.

Braithwaite, R., 1990b, 'More Than a Home for White Ants', *Australian Natural History*, vol. 4, no. 23, pp. 306–13.

Brunig, E. F., 1966, *Der Heidewald von Sarawak und Brunei. Eine Studie seiner Vegetation und Ökologie*, Habil-Schrift: University of Hamburg.

Groombridge, Brian (ed.), 1992, *Global Diversity: Status of the World's Living Resources*, London: Chapman and Hall.

Powell, J. 1978, *Mirrors of the New World: Images and Image Makers in the Settlement Process*, Canberra: Australian National University Press.

Whitmore, T. C., 1975, *Tropical Rainforests of the Far East*, Oxford: Clarendon Press.

24 Eating the future: Strangers in a strange land*

I find *The Future Eaters* a difficult book to review for a range of reasons, two of which are personal, others general. One is that I think that this is an important and an exciting book; I want to say that 'This book should be read by all Australians', but such rhetoric makes academics feel uncomfortable, since encomia seldom wear well over time. But I do rank it with the handful of books that help me in the unending struggle to learn who and where I am in Australia and the world – books like Bernard Smith's *European Vision and the South Pacific* (1960), and Joe Powell's *Mirrors of a New World* (1978), among others.

A second reason is that the breadth and complexity of the themes are beyond summary, although I shall have to attempt one shortly. A third is that some 'facts' and hypotheses are bound to be disputed, or, indeed, later shown to be wrong or misleading, but this is a task for all the professional experts whose work is so comprehensively synthesised and a task that many of them will doubtless address. Of the two personal ones, the first is that the author's academic qualifications in English and environmental science, combined with a concern to understand human behaviour from an ecological perspective, make it like meeting a member of your tiny ethnic minority in an alien land. The second personal one is pure chagrin, in the manuscript of my current book (that is, the book you are now reading), I give a sketch of my view of Australia as a place, which is not the cornucopian view that still seems to prevail. Tim Flannery does it better, pursuing most of the same themes, though happily from a mildly different perspective (this sketch became Chapter 23, The lie of the land, which I wrote before reading Flannery).

The subtitle of his book is *An Ecological History of the Australasian Lands and People* and the author describes himself as a 'historical ecologist' (p. 272). We are a species among others, subject to similar imperatives: the primary assumption is that popularised by Desmond Morris a decade or so ago in his books on animal behaviour. Our species is shown in comparative context: we are a predatory species,

*First published as 'Strangers in a Strange Land: An Essay Review of *The Future Eaters* by Timothy Flannery', *Meanjin*, vol. 54, no. 2, 1995, pp. 290–301.

and a particular kind of predator, a flexible one, able to switch from one resource to another, and thus unlike those predators that are restricted to a limited prey. In the latter case, a decline in prey numbers leads to population reduction in the predators, and there is thus a degree of self-regulation in the system. Not so with the flexible predator, who can exhaust one resource after another. This is our history, and thus we are future eaters. Our species in Australasia has consumed much of the resources on which its future depends.

This is not, however, the kind of doom and gloom we have had from the Club of Rome or *The Limits to Growth* (Meadows et al., 1972). If the conclusions are similar, we get there by a very different route. The main theme has, from the outset, some sub-themes that have been familiar for a decade or more, but others are startling. 'The first Australasians were . . . the very first humans to escape the straitjacket of coevolution' (Flannery, 1994, p. 155) is the key claim – and this was an event of major importance for all humanity, because 'it altered the course of evolution for our species'. Somewhere, perhaps about 60,000 years ago, our species made a great leap forward, from being 'just one uncommon omnivorous species among a plethora of other large mammals', all but lost 'among the great herds of mammals of their Afro-Eurasian homeland' to being the dominant species. What enabled the leap? 'The acquisition of language' is a popular answer. Flannery rejects it: 'a pharynx capable of producing speech had developed by 400,000 to 300,000 years ago – long before the great leap forward' (p. 156). Moreover, the people most likely to maintain very primitive languages, the long-isolated people of Australia and New Guinea, have languages quite as complex and sophisticated as the languages of Europe. If, however, language evolved only 60,000 years ago, 'it must have become universally adopted almost instantaneously in order for it to have achieved its present universal level of complexity' (p. 157).

Are there other explanations? Living in a tightly coevolved ecosystem, in which our species had evolved along with lions and tigers and zebra and gazelles, makes it hard to utilise new resources because there is always another, more specialised species already harvesting each resource efficiently. Thus it is difficult to take even a step forward, let alone a great leap. Moreover, there is a kind of unceasing arms race – if one organism takes even a small step forward, all the others affected by it, 'whether they be predator, prey or parasites – take a step also . . . there is often a zero sum game' (p. 159).

The release from coevolution

To make the leap, our species must have been suddenly released from the tight constraints of coevolution by leaving them behind, to have experienced for the first time 'conditions which allowed population to build up substantially', where sheer survival was not a ceaseless preoccupation, allowing a little freedom to experiment, to develop a ' "managerial" mentality', and where competition between various human groups was significant, impossible at low densities. 'Until 60,000 years ago, humans did not inhabit – and had never inhabited – such an environment. By 40,000 years ago, however, they were well established in over 10 million square kilometres of such habitat, for they had invaded Australasia' (p. 159).

The beach head was probably Lombok, east of Wallace's Line. Bali had tigers and leopards: Lombok had neither. There was a myriad new ecological opportunities. Humans could co-opt virtually every resource available to them. In Bali or Java, similar resources went towards maintaining thousands of other species, many of them big energy users, like elephants and rhinos. In Lombok, population build-up may have been quite rapid; so, it is argued, was the acquisition of new skills. As the resources acquired with least effort were used up, the flexible new predator devised ways of using other resources and so on to a point of resource decline, accompanied by population decline, eventually reaching an equilibrium at much lower levels of consumption and population – followed by emigration to the next land mass, where the cycle was repeated.

This scenario applies equally to Aboriginal, New Guinean, Maori and European invasions. They differ primarily in timing and in the character of the new environment, both of which are at least as significant as the technology of the invaders, often presented as the prime differentia. We are given a glimpse of a 'naive land', Lord Howe Island in 1788 – one that had never seen either humans or top predators – through the eyes of the great French navigator La Pérouse, followed by two vessels of the First Fleet. All of the birds were so tame they could be approached on the beach and simply knocked down. One observer writes: 'When I was in the woods amongst the birds I cd. not help picturing to myself the Golden Age as described by Ovid'. The birds never made the least attempt to fly away: 'The Pidgeons were also as tame as those already described & wd. sit upon the branches of trees till you might go & take them off with your hand'. So that is what they did: 'many hundreds of all the sorts mention'd above, together wt. many Parotts, Parroquettes, Magpies & other birds were caught & carried on board our ship' (p. 177). And so the Golden Age was brought to an end as soon as it was recorded.

Easy pickings: the Aborigines as future eaters

Australia was never so fertile, but 60,000 years ago it had an abundant fauna of large animals – the megafauna – and their extinction post-dates the arrival of the Aborigines. Flannery attributes the extinction to hunting, just as the Maori wiped out eight species of moa, beginning only 800 years ago. Flannery rejects the attribution to an increasing aridity, on the grounds that the megafauna disappeared everywhere at much the same time, from the semi-arid interior, the high plains of the Monaro, the fertile coastal valleys and the coastal rainforests. Increasing aridity would have left some refugia, especially given that some species were clearly already adapted to sub-arid conditions. He does not discuss the views of anthropologists such as Mulvaney that the extinction was primarily due to habitat changes, probably induced by the Aboriginal use of fire. This seems more probable to me than direct hunting, but it is a minor point, in that both are the result of human intervention.

The effects of fire

The fire story is fairly well known, but Flannery gives it a new twist. Between 60,000 and 40,000 years ago, the flora of much of Australia consisted of rainforest species; they should more precisely be called fire-sensitive species, since the rainfall was not appreciably greater than it is today. They include the southern beeches (*Nothofagus* species), the southern pines such as *Podocarpus* and *Araucaria*, tree-ferns and the like. Sediment cores carrying preserved pollen have been used to

234

establish these conclusions from a number of sites. Then there is a fairly rapid replacement by fire-promoting species such as the eucalypts and other sclerophylls. Large quantities of charcoal also began to appear in the sediment sequences, but this does not correspond with any major climate change. This is the familiar story. But the increase in fire intensity, according to Flannery, is not directly due to Aboriginal burning. There are enough natural causes of fire already. Aboriginal fires would have increased frequency, but actually reduced intensity – a hypothesis surely confirmed by the consequences of the cessations of Aboriginal burning over the last hundred years, which has resulted in fewer fires, of disastrous intensity (see colour illustration following p. 174).

Fire intensity is increased when the standing fuel load is increased. So how was it suddenly increased between 60,000 and 40,000 years ago? 'Through the accumulation of vegetation which would normally have been recycled through the guts of large herbivores' (p. 229) – the megafauna hunted to extinction by the Aborigines. We go to Serengeti for comparison: 'Imagine . . . if all of the large herbivores were removed . . . What would happen? Without doubt, the uneaten vegetation would build up dramatically. Given the right conditions, vast wildfires would rage' (p. 230).

It seems plausible, but of course one wants to argue, as with other of these hypotheses. It didn't happen in New Zealand when the moa went, although the Canterbury Plains always had some summer-dry grassland. Flannery has an answer for this. But even in Australia, it seems unlikely that huge amounts of fodder were left unused for long. It is a truism of ecology that no available niche is left unoccupied for long, and if the large herbivores disappeared, surely the small ones would have multiplied and diversified to take up the slack. One has to suppose that the evolution of a fire-promoting flora was faster than the evolution of the smaller herbivores. Equally, of course, one has to suppose that the large herbivores were nearly as helpless in the face of Aboriginal hunting as the moas were for the Maori or the birds of Lord Howe to the First Fleeters. But Australia already had a complex ecosystem with top predators such as the marsupial lion, *Thylacoleo*, and the Tasmanian tiger, *Thylacinus*. Can we assume that they were grossly inefficient? Tasmanian shepherds did not think that of the thylacine. However, the megafauna went. So did most of the rainforest, and fire became commonplace. Anyone who seeks to dispute Flannery's interpretations needs to find a better story. He applies it equally to the Aboriginal and to the European invasion. After rapid exploitation, a profound impact on the vegetation and thousands of years of nutrient loss through fire, the Aborigines at length reached an equilibrium in an impoverished land. Europeans began the process all over again: they found new ways of exploiting the remaining resources, and built up rapidly in population, at the same time reducing the long-term carrying capacity of their environment. They are future eaters, like the Aborigines before them.

European global future eating

Flannery, however, links all of the European expansion into the New World with the industrial transformations of the nineteenth and twentieth century. This is a second leap forward, but one paid for in precisely the same way: 'it was . . . fuelled by . . . the previously inaccessible but immensely rich soils of the Americas' (p. 163). There is an analogy with the great mineral wealth of Australia. Both are essentially a one-

235

off bonanza. He does not argue that we should not seize such opportunities, but rather, that they are rare and special events – so that we should think carefully about how we use the proceeds. Today's Australians eat them, spend them on consumables. The moral of the tale is that the first great leap forward was achieved at 'the cost of the destruction of whole biotas, while the second leap resulted in more destruction'. We are now in the middle of a third great leap, that of global information technology and of automation. 'We cannot afford to leap at such expense again' (p. 163).

However, a summary sketch gives a poor idea of the sweep and breadth of the argument of this book. The comparative scope is productive: the subject is more than Meganesia, as Australia, Tasmania and New Guinea are referred to; it is the fragment of Gondwana known to geologists as Tasmantis, which adds New Zealand, New Caledonia, and assorted outlying fragments such as Norfolk, Lord Howe, Kermadek and the Chatham Islands. Even Europe is served up: 'If we are to understand Australasian history properly, we must understand a little of the ecology of the Europeans. To do that, we must see Europe over 35,000 years of environmental change' (p. 303).

The key role of the Ice Age in Europe

As discussed in Chapter 23, Europe underwent massive glaciation, which left fresh nutrients from ground up rock, but the predominantly east–west mountains formed an insurmountable barrier for a biota attempting to retreat south before the cold. So she was pretty much laid bare. When the ice retreated, there was a rush north by plants that can colonise rapidly, invasive plants with many of the characteristics of the plants we know as weeds. The animals that recolonised were also those that were pre-adapted to disturbed environments. 'Such species are usually generalists with broad ecological niches.' They can breed and disperse rapidly, and can tolerate close human settlements. They are also profligate energy users, but this is not a problem in a nutrient-rich environment. 'Mobile, fertile and robust, Europe's life forms were purpose-made to inherit new lands' (p. 304). They have had good survival rates in Europe, despite the density of human settlement: 'of all mammal genera whose members exceed 44 kilograms in average adult body weight, Europe has lost only 29 per cent of those that were present there some 20,000 years ago' (p. 308), against the loss of 94 per cent of such animals in Australia.

The survival of these large mammals has been an immense advantage. The extreme example is the horse, which has contributed so much to food production, milling, transport, hunting, and warfare: 'horses became the unstoppable Sherman Tanks of ancient war' (p. 309). So 'the critical assets of a rich, weedy environment with a reliably seasonal if severe climate and retention of a variety of large mammal species' gave Europeans the edge, permitted them to expand to the four corners of the globe. The fate of these colonies and the fate of the people amongst whom they were established 'was largely determined by biology' (p. 311).

This search for biological and ecological explanation of historical events is pursued throughout the book, and yields some fascinating hypotheses. One is that in areas in which nutrients such as nitrates and phosphates are in plentiful supply, the species that are best at utilising these nutrients can out-compete all similar species, the prime example being ourselves. Where they abound, along with water and soil, we can destroy almost all other species through intensive agriculture, producing a monoculture. When the supply is less, such as in areas used for grazing, our ability to destroy competing species is less, although we are improving that ability. In the least productive environment, humans remain just another species among many, if they can exist there at all (p. 94).

New Guinea, New Caledonia, New Zealand

The accounts of New Zealand, and New Caledonia and New Guinea greatly enrich the story. The New Guinea comparison is especially enlightening, because both the New Guineans and Australians are sprung from the same stock, and were in contact with one another until 10,000 years ago. But environmental pressures and opportunities were very different, especially in the Highlands, where deep and fertile soils and good rainfall encouraged the New Guineans into agriculture at a very early date. 'Indeed, they appear to have been among the world's first agriculturalists. They are certainly amongst the world's most successful' (p. 292), supporting about 1614 people per square kilometre in the more prosperous highland valleys, 'the highest rural population density found anywhere on Earth' (p. 296). Mick Leahy's oft-quoted words on discovering the gardens of the Whagi Valley are repeated: 'The green garden patches were a delight to the eye, neat square beds of sweet potatoes growing luxuriantly in that rich soil, alternating with thriving patches of beans, cucumbers and sugar cane' (p. 293).

The conclusion to this chapter is important enough to quote in full:

> New Guinea, clearly, has major lessons for the interpretation of Australia's history. It shows that, given the right environment, people of Australoid stock are able to become highly successful agriculturalists. It also shows that, at a time when agriculture was just a distant glimmer on the south-western horizon for Europeans, New Guineans were already feeding themselves from their well-made gardens. It is also important because it shows that Australoids could develop societies that live in extraordinarily high densities. But most important of all, it shows that Aboriginal society was not primitive, nor was it the only option available to the people of Meganesia. (p. 298)

237

This may sound like the conventional liberalism of educated Australians today, but Flannery follows where his arguments lead him, with little regard for political correctness. He is withering in his scorn for the museums and lobby groups for giving over 'Aboriginal osteological material' (that is, bones) for reburial, since the bones are those of people who could not describe their lives in writing for their descendants, but through the study of the bones, living Aborigines can learn much about that past. 'As bleak as it is to be a future eater, it is as nothing compared with the tragedy of obliterating one's past' (p. 278). Much later in the book, multi-culturalism is seen as part of the cultural cringe, our failure 'to develop a strong vibrant and unique culture' of our own. Neither multiculturalism nor talk of becoming a part of Asia will help us to live comfortably in our own land; in fact 'these policies could destroy the slow process of adaptation between humans and their environment which is currently occurring in Australia' (pp. 394–5).

Management strategies

In its last pages, the book addresses management strategies. The main thesis stands without these pages, which are likely to provoke debate, and be seen as extremist by professional resource managers, but I find them refreshingly honest, and in accord with what many thoughtful Australians think privately, namely, that we have stuffed the place up, and are profoundly uncertain about what we should do next; or in Flannery's words 'the conservation challenge in the "new" lands is, in general, far more difficult' than is officially recognised (p. 388). This holds equally well for New Zealand, where the annual direct cost of one introduced vertebrate, the Australian possum, is $NZ32 million, and, when indirect costs are estimated, a total of $NZ2.00 per person per day (Rooney, 1995, p. 20).

Flannery gives us three choices: to keep our reserved lands as they are today; or as they were 200 years ago; or as they were 60,000 years ago! If we are aiming for biological diversity and ecological stability, the logic of this book requires us to consider the third possibility. 'Until around 60,000 years ago, Australia's ecosystems were fully self-sustaining. Then, vast extinctions devastated the entire continent'. In time, the Aborigines learned to manage 'the crippled ecosystems, preventing them from degenerating further. Now Europeans have arrived and forced the dis-continuance of that management' (p. 379). One consequence is the frequency of devastating wildfires, of which the 1994 Sydney fires are the most recent. Eric Rolls in *A Thousand Wild Acres* (1981) and Sylvia Hallam in *Fire and Hearth* (1979), among many others, have shown very clearly the great changes in the character of the so-called 'natural environment' over the last 200 years, changes that make the concept of 'wilderness' as held by some conservationists today demonstrably untenable. But going back to Aboriginal fire-management poses all sorts of practical problems (even in Kakadu, which is exceptionally well resourced, comparatively speaking, and capable of mounting experimental research programs into fire-management with CSIRO).

Bring on the Komodo dragon?

So Flannery looks at the alternative of going back 60,000 years! The need is to fill in the gaps, especially in bringing back the large herbivores and the large carnivores, all long extinct. For the carnivores, we could at least reintroduce the Tasmanian devil to the mainland, perhaps breed up the dingo again (although it is a recent arrival), and perhaps introduce the Komodo dragon to northern Australia, to keep the salt-water crocodile company. The popular press might easily make sensational headlines from such a proposal, but it is argued thoughtfully and cautiously. The Komodo dragon is the closest living relative of Australia's lost reptilian carnivores, and they

> have the potential to bring greater ecological stability to Australia's tropical regions. This stability could be achieved by helping to recreate a guild of carnivores that would regulate herbivore numbers enough to smooth out the cycle of boom and bust that currently characterises Australia's introduced placental megafauna. (p. 385)

The indigenous biota is also unstable: if you saw the images on television early in 1995 of emus piled up in their thousands on a fence line in Western Australia, driven from further inland by drought, you would recognise the scale of the problem. The Komodo dragon, along with the Tasmanian devil, might in time reduce the number of imported carnivores through competition for food and by preying on them. There are no foxes in Tasmania, despite several attempts to introduce them. Ridding the continent of foxes alone would make an enormous difference to the avifauna and small marsupials. More complex grazing and browsing patterns might even allow the expansion of fire-sensitive species and reduce the range of catastrophic wildfire.

A brief account of this thesis cannot do it justice, and is likely to be even more provocative than the original account, which is itself provocative. But I welcome it. It is a relief from the glib superficiality of chatter about 'sustainable development' in a continent that is now in a state of ecological instability.

Courageous, risk-taking books such as this are vulnerable, and of course I have points of criticism. There are statements that I see as overconfident, some inconsistencies, some errors. Nevertheless, Flannery generally plays fair by presenting the contrary evidence. For example, chapter 20, with the title 'Time Dwarfs', explores size reduction. Most of our existing native species, including fish, have been getting smaller. This is quite well attested; for example, the eastern grey kangaroo (*Macropus giganteus*) is some 13 per cent smaller than its ancestors of some 20,000 years ago. The red kangaroo, the wallabies, Tasmanian devils, all show the same trend; even koalas show it. Climate change and declining levels of nutrient availability are probably contributory, but Flannery also invokes hunting pressure. If it takes a long time to stalk a kangaroo and to kill it with the available weapons, you might as well stalk a big one. So selection for size operates. The wombats, however, do not seem to have shrunk, yet they are big marsupials, reaching 40 kilograms in weight. Perhaps their burrowing habit protected them, says Flannery, or our measuring system (largely based on teeth) is giving us wrong answers. Flannery is not happy with his tentative explanations, but at least he shares the problem with us (p. 216).

For apparent inconsistencies, one might consider the weight that is put on the El Niño Southern Oscillation (ENSO), the phenomenon behind the unpredictability of the Australian rainfall, varying from drought to flood. That and the low-nutrient availability are the major components of what is colourfully called 'racetrack Australia', on which 'high octane placentals seemed to burn out very early on, while the marsupials, with lower resting metabolic rates and thus lower energy needs, came in clear winners'. The monotremes took a poor second place, but at least finished the course, although they failed to do so in South America, where the placentals romped home (p. 74). But if the marsupials won the race in Australia because of their lower energy needs and their ability to sit out the bad times inflicted by ENSO, what about south-western Australia, which is little affected by ENSO and has at least comparatively reliable rainfall (or less unpredictable), yet the ecosystems of the south-west do not seem to differ significantly from those of eastern Australia.

Imperfections include errors that a good editor should have picked up. Flannery gives us things more important: freshness of language and great sweeping themes, clearly articulated. We get a well-informed, well-written, passionate book on themes of real importance to us all. So I will say it. This book should be read by all Australians. Since Flannery brings Australia centre-stage, we might also bring it to the attention of those lurking in the wings ('down-under' in the northern hemisphere).

239

References

Flannery, Tim, 1994, *The Future Eaters: An Ecological History of the Australasian Lands and People*, Sydney: Reed Books.

Hallam, Sylvia, 1979, *Fire and Hearth: A Study of Aboriginal Usage and European Usurpation in South-western Australia*, Canberra: Australian Institute of Aboriginal Studies.

Meadows, D. H., Meadows, D. L., Randers, J., and Behrens, W. W., 1972, *The Limits to Growth*, New York: Universe Books.

Powell, J. M., 1978, *Mirrors of the New World: Images and Image Makers in the Settlement Process*, Canberra: Australian National University Press.

Rolls, Eric, 1981, *A Million Wild Acres: 200 Years of Man and an Australian Forest*, Melbourne: Nelson.

Rooney, Derrick, 1995, *Press* (Christchurch), 8 March.

Smith, Bernard, 1960, *European Vision and the South Pacific*, Melbourne: Oxford University Press.

25 Felling the 'Groves of Life'

That the frequency of fire in Australia, which leads to continuous and progressive nutrient loss, is historically partly a function of long-term biological imbalance, is probable. The 'proposal' for rebalancing the ecosystem by introducing into the grasslands of northern Australia a high-level predator such as the Komodo dragon is worth considering from a narrowly ecological point of view, and wickedly provocative from a social one. It certainly dramatises the problem. This suggestion is for the grasslands of the north, where the dragon might make a plausible if colourful and difficult addition. But the really urgent fire problems are on the outskirts of Sydney and Melbourne. The release of a few top predators in the Blue Mountains and Dandenongs might meet with some resistance. The message is simple. This is a disturbed landscape, and the management options are both complex and inadequately imagined and explored.

The Future Eaters is written with the passion and immediacy that I now find Chapter 22 to lack. So is Geoff Park's book, *Nga Uruora*, mentioned briefly in the introduction to this section as communicating the author's sense of what New Zealanders have lost (Seddon, 1996). The whole book is about loss: its 384 pages ache with the feel of it as an amputee is said to ache with the pain of lost limbs. What is lost are the lowland forests of the well-watered, fertile plains of New Zealand, the 'Groves of Life' of his title. The narrative sets out to evoke the richness and beauty of what met the first Europeans, the relentless horror of its destruction and the banality, poverty and insecurity of the landscapes that have replaced it. It struggles to explicate the meanings of the loss and the acts that occasioned it, meanings that are intensely personal to the author as an individual, but also as a New Zealander, or one who is learning what it might be to be a New Zealander.

At least one of those meanings is unambiguous, and Captain FitzRoy, the master of the *Beagle* on which Charles Darwin visited New Zealand, got it in one: 'the Natives', he told the British House of Lords, 'live almost entirely upon the sea coast . . . the part of which is immediately around the Harbours, which is also the Part which would be of most Use to Settlers' (Park, 1995, p. 314). The European settlers took the best lands, as they did in Australia, displacing those who preceded them. In the case of New Zealand's mountainous islands, these were the coastal plains, which were at once by far the richest ecologically, the most densely populated, and the most limited resource. This is at the heart of today's New Zealand 'Maori problem', but it is also a significant part of her agricultural problem. The most productive pastures of New Zealand are rarely natural grasslands, especially in the North Island. They are established on drained and cleared rainforest, forest which built up porous, humus-rich soils that now undergo depletion. There are both parallels with and differences between the New Zealand and the Australian stories of settlement, one of the differences being the abundance of sclerophyll woodland and natural grassland in pastoral Australia, whereas the dairy farms and fat lamb pastures of New Zealand are generally the equivalent of the struggling dairy farms of the Dorrigo Plateau, the Atherton Tablelands and the Northern Rivers, similarly on cleared rainforests.

Now, the author says of one such farm in New Zealand, 'there is not a native plant to be seen. Once, before the hooves of the dairy economy, you could have plunged your hand into loose, rich, pliant humus. Now my feet are quickly clogged with the gouged, clay ground of a meaner soil' (p. 23).

The dominant trees of these lowland forests were kahikatea (*Dacrycarpus dacrydioides*). Kahikatea are ancient survivors from a world 'in which huge ammonites stalked the sea-floor and pterodactyls the air' (traces of kahikatea pollen and leaves have been identified by palynologists and palaeobotanists in rocks from the Jurassic, some 160–180 million years ago. 'Kahikatea persists from an old, swampy, worn-down, tropical archipelago, utterly different from the cool, young, mountainous New Zealand of today. You can find it in the hills, but it only prospers in the swamps. It would vanish without them' (p. 36).

These forests are the 'Groves of Life': the small fruit (koroi) are superabundant and highly nutritious. The birds ate them and flocked for miles to do so. 'Go into a lowland kahikatea forest in autumn when its koroi are ripening, lie under the towering trees listening to the cacophony of birds and the constant patter of the inedible bits hitting the leaves around you, and you'll know what "the groves of life" mean' (p. 15). The birds ate the koroi: the Maori ate the koroi and the birds. In 1841, in autumn, the botanist James Bidwill – for whom one of our araucaria is named – saw the ' "enormous kahikatea" ', 'loaded with their beautiful scarlet and black fruit', which 'formed a great part of the food of the natives during the season'. He counted sixty large baskets of them in one village, although their gathering must have been hazardous (p. 35). There is a Maori epigram to the point: *He toa piki ràkau kahikatea, he kai na te pakiaka* – the kahikatea climber is food for the roots.

Now they are all gone from the lowlands of the North Island, except for a few small remnants; there is a pocket in the South Island in the midst of suburban Christchurch, Riccarton Bush, there because of just one family who loved it; still a wonder and a sanctuary from its mind-numbing context. The cold wet West Coast of the South Island is their only major refuge, one in which they do not reach the former grandeur of those of Hauraki, south of Auckland. Captain Cook and other naval men thought they would be a great naval resource for masts and spars, but they proved not to be durable, and most of the warm forests were felled for butter boxes when the refrigerated export of butter became feasible from Australia and New Zealand in the 1880s.

The Groves of Life: *Araucaria bidwillii*, the bunya pine of south-eastern Queensland, is an emergent towering above the rainforest canopy in the Bunya Mountains. Like the kahikatea in New Zealand, it is one of the Gondwanan conifers, although not of the same family; and also like the kahikatea, its fruits were food and feast for the Aborigines, who congregated from afar in season. The land in the foreground was cleared for dairying. QUEENSLAND DEPARTMENT OF FORESTRY

New Zealanders have little feel for the reality of their loss, says Park. 'Mention of Nga Uruora today is like raising something from the dead' (p. 15). That is what he sets out to do, and succeeds in doing. It sits well with some recent Australian books, most notably *The Future Eaters*, and Tom Griffiths' *Hunters and Collectors* (1996). Geoff Park's Maori are future eaters, less so than Flannery's but Park's pakeha are more voracious: 'New Zealand's lowlands were found, possessed and gutted by a *foreign* culture' (p. 307); 'True "future eaters" . . . we have exploited these islands' richest ecosystems with all the violence that modern science and technology could summon' (p. 332).

Hunters and Collectors: The Antiquarian Imagination in Australia (Griffiths, 1996) is a fine companion piece. The two books might share Griffiths' conclusion and his epigraph, the latter a quotation from Judith Wright: 'These two strands – the love of the land we have invaded, and the guilt of the invasion – have become part of me. It is a haunted country'. The conclusion: 'History – that stubbornly contextual and relativist craft – may be the tool that enables us to grope for a conservation ethic that is social as well as ecological' (p. 277). Geoff Park is an ecologist, Griffiths a historian, so their pathways to a similar conclusion differ, as do their context, but the comparisons are the more illuminating: the Maori past; the Maori presence; the Aboriginal past; the Aboriginal presence; and in both, the presence of the past – 'At times, the past forces itself shockingly upon the present, like an intrusion across a geological fault line' (Griffiths, p. 278) – and the two lands are so different, one so young, the other, so old: 'To explore Australian space was to plumb global time' (Griffiths, p. 9).

For me the overwhelming sense of loss that resonates through *Nga Uruora* has come, before the experience of this powerful book, not in New Zealand, but in Natal and in south-western Australia. In Natal, I was given a copy, newly reprinted, of *Travels in Southern Africa (from 1838 to 1844)* by a young Frenchman, Adulphe Delegorgue. It is a beautiful book, brimming with the enthusiasm and vitality of a young natural historian: 'loud in his protestations of dedication to science and love of natural beauty, Delegorgue rushes over the Southern African landscape, systematically destroying as much of its fauna as possible', as the editor acidulously remarks (p. xix). But what a landscape, magnificently alive with great trees, birds, lions, leopards, elephants, hippopotamus, the whole zoo, before, around, behind him, travelling north through coastal Natal where today there is one of the world's most monotonous and greedy monocultures: mile after mile after dreary mile of sugar cane, with little else to be seen, hardly a tree. In south-western Australia, the plains around Esperance and Ravensthorpe were cleared only after World War II, and the ecological poverty of the human settlements is an unbearable contrast with the exquisite diversity of the natural environment in the adjoining Fitzgerald River National Park.

Geoff Park often communicates his sense of loss by superimposing two views of the 'same' landscape, differing only in the time of the portrait. He does so in describing Te Mome stream, now a part of Wellington, the national capital. In 1840, one of the first colonists, Edward Hopper, found himself a beautiful spot: shaded by a Karaka grove where

> 'our dwellings appear to be built in a shrubbery ... trees of the richest foliage ... bearing a fruit the natives are very fond of.' ... Here in this place of 'enchanting scenery' and no wild beasts, three-to-six-pound fish landed every time a line was thrown in the river, and in the forest were parrots and pigeons 'twice the size of English Pigeons'.

But by 1990, the drainage board is having trouble with the stream, and where Hopper landed farming gear on the fertile alluvial soil to begin clearing

> 'the dense forest which has enriched the soil for ages', a mid-week match is on at the central Croquet Club. In the muddy, tidal mess alongside, all that remains of the river, a white-faced heron tracks along the film of water which slowly creeps over the urban debris of car wheels and bottles ... Day-glo graffitists have left 'Master Flash Rasta' on the corrugated iron. (p. 112)

The great diversity of birds has gone. 'They left land that sustained one of New Zealand's richest life support systems, in the control of a tiny biotic band, the industrial world's ubiquitous scavengers and perpetual pioneers – rats and mice, flies, gorse, willow, broom, oxalis and blackberry, sparrow and seagulls, and humans' (p. 112).

Oh brave New World ...

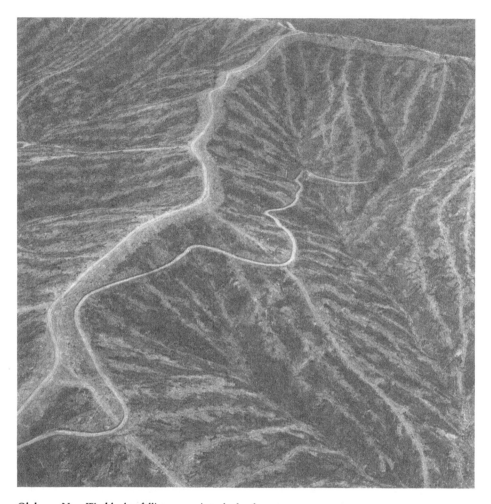

Oh brave New World: clearfelling overprints the land. NEVILLE ROSENFELD

References

Delegorgue, Adulphe, (1847), *Travels in Southern Africa (from 1838 to 1844)*, introduced and indexed by Stephanie J. Alexander and Colin De B. Webb, Pietermaritzburg: University of Natal Press.

Park, Geoff, 1995, *Nga Uruora: The Groves of Life*, Wellington: Victoria University Press.

Griffiths, Tom, 1996, *Hunters and Collectors: The Antiquarian Imagination in Australia*, Cambridge University Press.

Seddon, G., 1996, 'Oh Brave New World: Review of Geoff Park', *Meanjin*, vol. 55, no. 3, pp. 395–409.

Coda:
Learning to be at home:
'and then came Venice'

'And then came Venice.' In E. M. Forster's novel *A Passage to India* (1924), one of the key characters is on his way home, by ship, through the Suez Canal. Cyril Fielding is a minor civil servant in India, the schoolmaster in charge of the government college. He is a tolerant and reasonable man. Unlike most of his colleagues, he struggles to relate to India, even to make Indian friends, but in the end, it is all too much for him. He at last admits it to himself as he returns to what he sees as his own world:

> Crete welcomed him next, with the long snowy ridge of its mountains, and then came Venice. As he landed on the piazetta a cup of beauty was lifted to his lips, and he drank with a sense of disloyalty. The buildings of Venice, like the mountains of Crete and the fields of Egypt, stood in the right place, whereas in poor India everything was placed wrong. He had forgotten the beauty of form among idol temples and lumpy hills; indeed, without form, how can there be beauty? Form stammered here and there in a mosque, became rigid through nervousness even, but oh these Italian churches! San Giorgio standing on the island which could scarcely have risen from the waves without it, the Salute holding the entrance of a canal which, but for it, would not be the Grand Canal! In the old undergraduate days he had wrapped himself up in the many coloured blanket of St. Mark's, but something more precious than mosaics and marbles was offered to him now: the harmony between the works of man and the earth that upholds them, the civilisation that has escaped muddle, the spirit in a reasonable form, with flesh and blood subsisting. Writing picture postcards to his Indian friends, he felt that all of them would miss the joys he experienced now, the joys of form, and that this constituted a serious barrier. They would see the sumptuousness of Venice, not its shape, and though Venice was not Europe, it was part of the Mediterranean harmony. The Mediterranean is the human norm. When men leave that exquisite lake, whether through the Bosphorus or the Pillars of Hercules, they approach the monstrous and extraordinary; and the southern exit leads to the strangest experience of all. (p. 275)

I find this passage very moving. It illustrates many of the themes of this book. First, the power of words. Forster writes so well that the immediate effect is incantatory: it imposes its mood on any reader who shares a substantial part of Forster's educational background. Fielding is not, of course, the whole of Forster, who has another voice in the novel, Mrs Moore, who speaks for his intuitive intelligence. Fielding is only the rational Cambridge don voice of Forster, but it is he who is speaking here. The reflections suggest how Forster was enriched and crippled by Cambridge; immensely enriched by that imaginative possession of a human past that begins with the Greeks and embraced the Italian Renaissance; crippled by a limiting Eurocentrism, fully apprehending one world only by rejecting another.

My own first Passage to England, many years ago, was by sea, through the Suez Canal into 'that exquisite lake', Homer's wine-dark sea, and I too was welcomed by Crete and the long snowy ridge of its mountains. Our ship did not go to Venice; we sailed through the Straits of Messina, Scylla on one side, Charybdis on the other. Then on through the Straits of Bonifaccio between Sardinia and Corsica, Napoleon's birthplace, to Marseilles, once Massilia, an Ionian colony, founded by the Phocaeans six hundred years before Christ; later a Roman port serving Provincia, Rome's first province, now Provence. This was the world of my schooling, yet I was experiencing it for the first time; the world where I was born and have lived most of my life, I have had to learn for myself.

For the Suez Canal runs both ways, and the southern exit, leading to 'the strangest experience of all', takes us past India to Australia, 'where monsters be' and 'men do stand upon their heads' (and are torn by the inescapable tension of being Australian). Forster's literary skill lies in part in the way in which Fielding's emotional state is projected on to the landscape. The muddle is not India, but the social and political circumstances in which he had been caught up, and from which he had now escaped, without present responsibilities. Hence the euphoria, the sense of belonging and of harmony with his surroundings. In Tom Griffiths' lovely phrase, he had 'carved a home for the heart' from the wilderness of this world – as we all must.

Is there also objective truth in the passage, as well as subjective truth? Are the Indian hills in fact lumpy, lacking beauty of form? That is a harder question. My answer is 'yes and no'. I am reminded of my first response to the landscape around Perth: the trees seemed misshapen, 'everything was placed wrong'. I had to learn to look differently, and it took time, for at first I was *unsettled*. The concepts of harmony and proportion and composition are, for Anglo-Australians like myself, deeply embedded in European culture, as several of these essays show (for example, Chapter 8, Figures in the landscape): even those basic perceptions of foreground, middle ground and background are cultural constructs, and they are not operable in rainforest, where there is only foreground, or the Simpson Desert, where there is no middle ground. Perspective, roughly speaking, was a Renaissance invention, one that was developed in the clear light of the Mediterranean. A year ago, teaching a graduate class at an east coast American University, I asked the students how far was a long view. 'Sixty miles', said a boy from New Mexico. 'Half a mile on a good day if you are looking out to sea', said a girl from Boston.

Yet surely Forster's praise of the Venetian churches is just; of 'San Giorgio standing on the island which could scarcely have risen from the waves without it'. 'That harmony between the works of man and the earth that upholds them' is the lesson we struggle to learn. It is a continuing struggle, and a universal one, as critical in the Mediterranean as in Australia. The gates of Eden are firmly barred against us, and guarded with a flaming sword. There is no Golden Age, either fore or aft, no Shangri La, no El Dorado to be stumbled upon, no Bali-hai somewhere in the blue Pacific, only Mururoa. These are the enfeebling dreams. If there is life on Mars, it is not likely to have much to say to us, nor to enrich our lives. *This* is our garden of earthly delights. The earth is home. If we are at war with it, it is a war we cannot win; better to think of it as our partner, for richer or poorer, in sickness and in health, 'til death us do part.

References
Forster, E. M., (1924) 1954, *A Passage to India*, Harmondsworth: Penguin.

Index

abscission   80
Acacia rostellifera   173
Adams, Ansel   9
Adelaide Hills, SA   182
Albion House, Augusta   83, 87
algae   194
Alger, Eugène   88
Alice in Wonderland   126
alienation   21
allegiances   132
ambivalence   137, 177
Amelanchier canadensis   169
amethystine python (*Liasia amethystinus*)   94
Anatolian Plateau   178
Ancona   99
Anemone hupehensis   165
angiosperm
    evolution   75
    taxonomy   78
animist survivals   17, 119
ammonoids   74
anthropocentrism   16
anthropomorphs   121, 125
Antipodean, the   73
Antipodes   66
appropriation   106
Arago, Jacques   3
Araucaria spp   165
    A. araucana   165
    A. heterophylla   165
    A. bidwillii   243
arboriphobia   166
Arbutus spp
    A. andrachne   165
    A. X andrachnoides   165
    A. unedo   165, 166
Arcadia   176, 180–1
Arcadian landscape   145, 176
Arnhem Highway   xvii
Attenborough, David   17
Austen, Jane   xiii
Australian Alps   110
Avestic (Old Persian)   179
Ayers Rock   9, 223

'Babyspeak'   194
backyard   153
    rural origins of   158
Ballarat   109, 138
Banks, Sir Joseph   75

Banksia coccinea   18
Banksia praemorsa   164, 172
Banksia Men   122
Bare-Arse Creek   26
Barrenjoey Peninsula   117
Barrett, Bernard   157
basil   154
Bauhaus   140
Beagle   171
Bevan, Aneurin   190
Berkeley, Bishop George   10
bidgee-widgee (*Acmena anserinifolia*)   12
Blainey, Geoffrey   9
Blake, William   7, 11
Blinky Bill   119, 120
Bogong High Plains   110
Bonnet, Penelope   71
boundaries
    attitudes to   146
    maintenance   170
    national   133
Bowral, NSW   182
Boyd, Robyn   116
Braudel, Fernand   xiii, xv
Brooks, John   184
Browne, Sir Thomas   8
brush box (*Tristaniopsis conferta*)   92
buffalo grass   166
bush, tamed   87
Butler, Samuel   xi

Callitris preissii   173
Calocephalus brownii   174
Capability Brown   112, 179, 203
Cape Range   99
Carey, Peter   23
Carter, Paul   21, 24
Castellorizzo, Greece   101
cauliflory   93
Cephalotaceae   79
cesspits   157
cesspool   189
chains, links, perches   151
Champs-Elysées   142
Childe, Gordon   75
Chile   165
China   138, 146, 165
'choice' plants   169
chook-house   153
Church of St John, Camden, NSW   115

249

Ciccarone, Julia   xii
Claremont, WA   165
Clematis armandii   169
Clematis nepalensis   169
Club of Rome   187
coal   77, 81
coastal planning   xvi
Cobbett, William   8
Cockies Tongue (*Templetonia retusa*)   173
collage   94–6
Collingwood, Vic   157
Comalco   102
consumerism   162, 183
contemptus mundi   7
contrived disorder   18
controlled idealism   xvi
convict background   50, 161
Corner, Professor E. J. H.   80
Cottesloe, WA   139
Cronon, William   22
Cruden Farm, Langwarrin, Vic   180
Crusoe, Robinson   xv
Cuddlepie   17, 107, 119–26
cultural identity   136
cultural landscapes   116

Daintree, the   93
Dale, Lieutenant Robert   23
Dalmatian meadow   169
Dampier, William   30
Darling Ranges   110
Darwin, Charles   171, 192
density, urban   156
Descartes, René   xiii
design
   environmental   66, 112, 136
   'water-sensitive'   147
Design with Nature   20, 116
Devils Backbone, The   25
Dickens, Charles   149
diesel   xvii
'discovery'   73
dispossession   190
disturbance   13
'*divorce de convenance*'   132
DPs (displaced persons)   52, 140
Dr Who   128
drawbridge mentality   150
dreams of place   90
dual allegiance   xiv
Dublomb, Charles   30
dunes, crescentic   29
dystopia   177

ecospeak   22
Eden   97, 147, 176
Egerton Warburton, Colonel Peter   30
Ehrlich, Paul and Anne   190
Eliot, T. S.   120

Elliot, Brian   92, 111
emblematic functions   84
Embothrium coccineum   165
embryonic diapause   13
emotionalism   189
Enclosures, the   180
Engels, Frederick   130
environment
   Anglicisation of   84
   coping with   21
   disturbed   182
epiphytes   94
Erewhon   xi
Erythrina cf. *indica*   171
Eucalyptus spp
   E. botryoides   23
   E. calophylla   122
   E. camaldulensis   137
   E. citriodora   166
   E. erythrocorys   170
   E. ficifolia   117, 122
   E. gummifera   122
   E. haemastoma   123
   E. maculata   117, 122
   E. marginata see jarrah   117, 122
   E. niphophila   23
   E. rudis   137
   E. todtiana   123
Eurocentrism   12, 73–82
Europe's 'geographic unconscious'   xiii
Exmouth   100

fan palms (*Licuala* spp)   93
Fairlie, Henry   194
Fear the hose   147
featurism   114
'fertile crescent'   138
Fitzroy Crossing, WA   138
Fletcher Jones, Warrnambool, Vic   166
Fleurieu Peninsula, SA   117
Flinders, Matthew   88, 139
Flora Tasmaniae   74
Foigny, Gabriel de   xi, 182
Forrest, Sir John   86
Fortunes of Richard Mahony, The   xiv
Foucault, Michel   10
frangipani   172
freesias   173
Fremantle, WA   165, 170–5
Freycinet, Commander Louis de   30
Freycinet, Mme   30
fruit fly   182

Gaia theory   18
gallows humour   190
garage, the   159
garden(s)
   'bush'   145
   choosing a style   172

of earthly delights   179
of Eden   97, 147, 176
as enclosures   146
as exclosures   146
as forest clearing   145, 166, 168
as heath   145
'native'   145
as oasis   145
as Paradise   147
as tools of imperial power   182
vegetable   154
walled   179
Gardner, Peter   25
Gastrolobium calycinum   76
Gauguin, Paul   181
genius loci   106, 113–18, 147
Geoffrey Hamlyn   22
Germania   178
Gibbs, May   17, 107
Gill, Don   71
Giorgi, Piero   13
global supermarket   113
global village   127
Golden Age   112
Golden Mean   138
Gondwanan origins   176
Gould, Stephen Jay   12
Grampians, Vic   71
grasstrees   15
Great Apes   186
Great Chain of Being   8, 192
Great Expectations   149
Grey, George   30
gumnut babies   122

Hahndorf, SA   151
Hallet Cove, SA   76
Hands Off Everything Society   187
Hardy, Thomas   17
Harvey, WA   121
Hasluck, Alexandra   120
Haussman, Baron von   142
Hawkesbury Sandstone   114, 168
Hawthorn, Vic   165, 168
Heames, Rene   121
heathen   15
heraldic beings   27
heritage   187
hermaphrodites   xii
Herrick, Robert   19
Heysen, Hans   88, 118
Hills Clothes Hoist   159
Hoare, Michael   73
Hoddle Street, Melbourne, Vic   151
Hooker, J. D.   74
Hope, Alec   71
Hough, Michael   20
houris   178
Hume Freeway, Vic   115

hydrology   116

'idiocy of rural life'   130
Indian Hawthorn   174

Jabiru, NT   xvii
Jackson, J. B.   9
Jacquards, the   189
James, Henry   xiv
jarrah (*Eucalyptus marginata*)   110, 138
John of Gaunt   182
Johnson, Dr Sam   11
Journeyings   101
jungle edge   90
Jussieu, de   79

Kailis family   101
Kakadu National Park, NT   xvii
Kalbarri oysters   30
karst landscape   169
Keating, Paul   182
keeping up appearances   161
Kelly, J. O.   121
Kew Gardens, UK   74
Kingia australis   15
Kings Park, WA   21
Kingsley, Henry   22
Kosciusko National Park, NSW   187

la terre Australe   xi
landscape
   cultural   116
   pastoral   145
   possessing   21
   structure of   111
landscape architect   97
landskip   97
Lapageria rosea (Chilean bellflower)   169
Laportea sp.   70
lawyer vine   93
learning to be at home   xiv, 98, 119, 120, 138, 246
Leichhardt, NSW   154
lerp   9
Lévi-Strauss, Claude   26
lillypilly   92, 93
limestone, love of   167
Linnaean system   79
Linum purpurea   175
Little People, the   119
Locke, John   10
Loudon's nursery, UK   24
long barrows of Mona Vale, NSW   115
Lorrain, Claude   87
Los Angeles   156, 187, 195
Louvre   141
Lowenthal, David   9
loyalty   132
lusus naturae   12

Macedon, Vic   182
Mackellar, Dorothea   92, 118
Mackenzie, Bruce   117
Maddocks, Cheryl   19

Magnolia liliflora   165
maize   16
male domain   155
marri (*Eucalyptus calophylla*)   122
marsupials   12
Marx, Burle   146
Marx, Karl   130
McCalman, Janet   101
McHarg, Ian   20
Mead, Margaret   176
Meekatharra, WA   25
megafauna   13, 215, 227, 234
meiosis   21
Melaleuca spp
    M. *cardiophylla*   167
    M. *hugelii*   167
    M. *nesophila*   167
    M. *pubescens*   173
    M. *pulchella*   167
    M. *violacea*   167
Melbourne's private schools   101
Melia azedarach   92
Mendel, Gregor   193
Mildura, Vic   109, 137, 154
Mitchell, Major T. L.   76
Molloy, Georgiana   120
monofluoracetic acid   76
moral arena   193
More, Sir Thomas   176
Moreton Bay fig   92, 140, 170
Mornington Peninsula   141
Mt Buninyong   109
Mt Eliza, Kings Park, WA   83–8, 110
Mt Warrenheip, Vic   109
Muir, John   9
mulberry   170
mullock heap   109
multicultural Australia   136
Murray River   137
Muswellbrook, NSW   70
My Country   92

naming strategies   16, 120
Natal   11
national boundaries   133
national parks   xvii, 18, 187
natural–supernatural   7
natural–unnatural   7, 9
natural–human   7, 10
natural–artificial   7
'Neolithic Revolution'   75
Neutral Bay, NSW   121
New Zealand   181
new chum   22
Nimbin, NSW   129
Ningaloo Reef, WA   99

Noble Savage   178, 181
Nolan, Sydney   111
Norfolk pines   139, 165
North West Cape, WA   99

oil
    discovery   100
    Middle Eastern   129
Oleander (*Nerium oleander*)   174
olive   170
Open Garden Scheme   179
order–disorder   18–19, 146
Oscar and Lucinda   23
Ottawa, Canada   20
outhouse   155

Paddington, NSW   156, 168
Paddington Society   133
paesaggio   97
Palladio, Andrea   xiii
Pan   120
Paradise   176
    earthly   179
    garden as   147
    heavenly   179
parish pump   128
Park, Geoff   181
Parthenon   141
passion for neatness   161, 166
Passmore, John   xvii, 188
past, as tourist destination   106
pastoral
    economy   179, 180
    idyll   194
    landscape   145
pathetic fallacy, the   16
patriarchal rhetoric   22
Pei, I. M.   141
Pennisetum alopecuroides   173
pepper tree (*Schinus molle*)   117, 153
permaculture   155
Permian coal flora   77
Permian glaciation   77
Perth, WA   21, 121
pig-keeping   159
pitcher plant   79
Pitjantjara   120
placelessness   139
platypus   73
Player, Ian   11
Pleistocene Period   76
Pliestocene   xiii
Podocarpus elata   166
poinciana   113, 172
Pooh Bear   119
Port Fairy, Vic   155
Port Moresby, PNG   156
possessing the landscape   21
possession, taking   181
Powell, J. M.   74
power lines   xvi

Preiss, Ludwig   76
primitive affluence, concept of   180
pruning   160
pseudo-contemporaneity   128

psychic homeland   119
public open space   163

quarter-acre block   151
Queensland umbrella tree (*Schefflera actinophylla*)   174
Quercus spp
   Q. agrifolia   174
   Q. canariensis   174
   Q. ilex   174
   Q. suber   174

Raalte, Henry van   88
rank, distinctions of   178
Raphiolepis umbellata   174
regional accents   127
regionalism, rebirth of   127
Reich, Charles   194
Relph, E.   139
Repton, Humphrey   179
resident action groups   133
resources, non-renewable and renewable   187
Revivalists   189, 191–4
rhetorical modes   189
Richardson, Henry Handel   xiv
Richmond, Vic   154, 165, 169–70
river red gum   137
Roberts, Tom   87, 118
Robin Hood   178
Rocks, Sydney, NSW   156
Romantic Revival   8
Romneya coulteri   175
Rosaceae   12, 79
Rose series (postcards)   86, 88
Rottnest, WA   140
Rottnest Island tea-tree   173
Rousseau, Henri   90, 181
Rousseau, Jean-Jacques   181

Sackville-West, Vita   145
Sadeur, Jacques   xi, 181
Saint Francis of Assisi   192, 196
Samson, Fred   170
scale of attention   111, 138
Schama, Simon   96, 178
sclerophyll flora   74, 78
Scotch College   168
scribbly gum (*Eucalyptus haemastoma*)   123
sea-floor spreading   81
self-maintaining systems   116
Selfish Gene, the   18
sense of place   105, 107, 109–12, 136, 167
service yard   159
settlers, German   71
sexual ambiguity   178
sexuality, innocent   178

Shark Bay   30, 101
Sierra Club   182
Sissinghurst, UK   145
skillion roofs   115
Snape, Diana   117
Snugglepot   23, 119–26
Snugglepot and Cuddlepie   121, 123
soil conservation   xvi
South Perth, WA   121
southern beeches   75, 93
space, abundance of   100, 158
space
   display   160
   public open   163
spaceship Earth   128
Species Plantarum   79
specificity   139
Spyridium globulosum   167
Stannage, Tom   157
Stirling Range National Park, WA   18
Stone, A. H.   84
structure of landscape   111
Sturt, John   76
sugar gums   71
survey, cadastral   150
Suzhou, China   146
swamp cypress   93
Swan River Colony   21
Swan River cypress   173
Sydney–Melbourne   131
symmetry–asymmetry   146

Tacitus   178
Tannahill, R.   16
taxonomy of the angiosperms   78
technology
   high   100
   homogenising   114, 127, 139
teleological language   18
Templetonia retusa   167, 173
territorio   97
Theocritean poetry   180
Thomas, Keith   26
Thoreau, Henry David   22
Toffler, Alvin   129
Tomasetti, Glen   155, 160
tomatoes   16
Toowoomba, Qld   182
Totterdell, Colin   93, 94
Tower Hill, Warrnambool, Vic   116
Turner, J. M.   88

Uluru *see* Ayers Rock
unattainable, attitudes towards the   177
University of Melbourne   168
University of New South Wales   168
'unspoiled'   191
urban amenity   134
urban planning and design   xvi
Urban Revolution   75
Utopia   xi, 176

Index

Vallance, Tom   77
Venetian Republic   20
ventriloquism   145
Vergil   139
Victoria, Queen   85
Walsh, James   84
Walters, Max   79
Washing Machine, The   25
waste disposal   183
water
    conservation   146
    management   xvi
water-sensitive design   147
Webb, Dr Len   93
weeds, invasion   16, 182
Wemmick   149
Werner, Abraham Gottlob   76
Western Australian Petroleum   138
White City   123
white-tailed kingfisher (*Tanysiptera sylvia*)   94

Whitman, Walt   8
'*Whose* place?'   107, 136
Wilde, Oscar   xvii, 15
wilderness   22
Wilkinson, Jennifer   169
Williams, Fred   88, 139
Wimmera, Vic   70
Winston, Denis   158
Woldendorp, Richard   28
Wordsworth, William   8, 139
Wright, Judith   92
Wupperthal, Germany   130

Xanthorrhoea preissii   15
Xenophon   179

Yallourn   81
York Road poison   76, 80, 120
Yosemite   9

Printed in the United States
By Bookmasters